Multimedia Applications Development

Intel/McGraw-Hill Series

INTEL • *386SX Microprocessor Programming Reference Manual*

INTEL • *i486 Microprocessor Programming Reference Manual*

MARGOLIS • *i860 Microprocessor Architecture*

LUTHER • *Digital Video in the PC Environment*

RAGSDALE • *Parallel Programming*

Forthcoming volumes

BABBAR AND STEUAK • *The Official Intel 386SL Portable Computer Book*

GUREWICH • *Communications Systems*

STECIAK • *Embedded Systems*

Multimedia Applications Development

Using DVI® Technology

Mark J. Bunzel

Sandra K. Morris

McGraw-Hill, Inc.

New York St. Louis San Francisco Auckland Bogotá
Caracas Lisbon London Madrid Mexico Milan
Montreal New Delhi Paris San Juan São Paulo
Singapore Sydney Tokyo Toronto

Library of Congress Cataloging-in-Publication Data

Bunzel, Mark J.
Multimedia applications development : using DVI technology / Mark J. Bunzel, Sandra K. Morris : foreword by Nick Arnett.
p. cm.—(Intel/McGraw-Hill series)
Includes bibliographical references and index.
ISBN 0-07-043297-X :
1. Interactive video—Computer programs. 2. Digital television.
3. Computer software—Development. I. Morris, Sandra K.
II. Title. III. Series.
TK6687.B86 1992
621.39′87—dc20 91-35288
CIP

1 2 3 4 5 6 7 8 9 0 DOC / DOC 9 7 6 5 4 3 2 1

ISBN 0-07-043297-X

The sponsoring editor for this book was Daniel A. Gonneau, and the production supervisor was Suzanne W. Babeuf. It was set in Century Schoolbook by North Market Street Graphics.

Printed and bound by R. R. Donnelley & Sons Company.

Contents

List of Figures

NOTE: All figure titles followed by an asterisk (*) appear in the color insert which follows text page 124.

CHAPTER 1

CHAPTER 2

CHAPTER 3

CHAPTER 4

CHAPTER 5

CHAPTER 6

CHAPTER 7

CHAPTER 8

CHAPTER 9

CHAPTER 10

CHAPTER 11

CHAPTER 12

CHAPTER 13

CHAPTER 14

Foreword

The marriage of computing power and television is one of the most important developments of the technology revolution. It has the potential to strike a balance between lethargic "couch-potato" television watching and the frantic interactivity of computer games and applications. Somewhere between beer and caffeine, there's a middle ground.

Computers are powerful productivity tools, but tools alone are boring, especially if a lot of expertise is needed to operate them. The old joke about how computers are hard to understand goes, "What's the difference between a computer salesman and a used car salesman? The used car salesman *knows* he's lying."

Television is easy to use and quite attractive, even addictive, but offers little more than a barrage of sensory stimulation. In a *Calvin and Hobbes* cartoon, Calvin bows before his television, saying, "O great altar of passive entertainment, bestow upon me thy discordant images at such speed as to render linear thought impossible!"

Computers are hard to use, but becoming easier through well-thought-out interfaces. Television is easy to use, but becoming harder through the rise of cable, satellites, and the incomprehensible seven-day, 14-event VCR timer (and computer people thought the "C>" prompt was hard!).

Today, with tools such as Intel's DVI® technology, people like you are marrying television's uncanny ability to attract and hold our attention with the computer's ability to start and stop at will, or to switch among a variety of media. The time is right.

As a motivator and attention-getter, television has been a fabulous tool for advertising over the last 30 years. But television's success is wearing out. Our attention is drifting, partly because more channels are available, but more likely because we're reaching the limits of sensory stimulation. Scene changes in television commercials can't become much shorter without passing the boundary of human perception. Something new is needed. It's not High-Definition Television (HDTV), which is just bigger, sharper lights and noise. Bigger, stupid televisions would be a Band-Aid. Smarter televisions are the next step.

While television has been an obvious influence on life in the past 30 years, computers have been an equally powerful, though much less visible, shaper of the recent decades. The plummeting cost of information management has led to an amazing variety of products and services that depend on data processing.

People are overwhelmed by the flood of publications, phone calls, television stations, direct mail, videotapes, conferences, and other goods and services screaming for attention—virtually all of them managed by, and many produced by, computers.

GAINING CONTROL

Despite the impact of television and computers—certainly the most profound technological influences in today's world—the average person has had little control of either one. *Programming,* which means something different in television and computing, has been out of the reach of the ordinary person.

The computer's ability to handle a variety of media—text, graphics, audio, video, paper, and telecommunications—offers a myriad of new opportunities to create products that allow ordinary people to cope with the sensory and factual deluges that threaten to overwhelm us. It won't happen overnight. The cost of these systems means that business will be the first to adopt them. There are four major applications in which interactive multimedia technologies will gain ground: learning, persuasion, documentation, and visualization.

LEARNING

The power of interactive video training and education is already known, especially when it is impractical for logistical, cost, safety, or other reasons to have a "live" teacher. Interactive video proceeds at the students' pace. It is always available, flexible to the student's schedule. It can simulate dangerous or expensive procedures such as aircraft operations and complex manufacturing equipment.

Yet this is only the start. The computer's ability to *manage* audiovisual resources is much more important than its ability to substitute for the teacher. There are numerous failures of interactive video in classrooms, simply because instead of helping the teacher do his or her job better, the courseware tried to impose its own curriculum, trying to replace the teacher.

The same tools that allow the teacher to manage audiovisual information, in the hands of students can bring them into the modern world of audiovisual communications as *author,* instead of remaining simply couch-potato recipients. With luck, this means that they will be far more aware and critical of the tricks and techniques that the electronic media employ to persuade and shape opinions.

PERSUASION

The power of audiovisual information to persuade (motivate, sell, indoctrinate—choose your word) is being put to work in sales and marketing, proposal preparation, and other persuasive business applications.

There's been limited success with kiosks and other multimedia productions that attempt to take over the sales representative's job. But the same lesson learned by education applies to sales and marketing—instead of trying to replace the salespeople, the best applications help them do their job better. The software teaches them about the products and helps them manage information so that they can serve customers better. In an age of increasingly customized products, interactive technology offers a way to cope with the overwhelming number of choices available.

At a personal communications level, even simple voice annotation of computer documents can become a powerful tool to put the right "spin" on them.

For example, thousands of computerized spreadsheets are sent from place to place every day, containing proposals, budgets, business plans, operating results, and so forth. They're boring. Worse, the people who send them really have no good way to draw attention to key points. Pages of text, explaining the numbers, may accompany the spreadsheets, but anyone who thinks those really have an impact probably also thinks that people actually read software manuals. They don't.

Voice annotation, ideally combined with software that records the screen image, lets the sender attach significance to certain parts, much better than is possible with text annotation. That's a huge step in communications. It's not simply that the computer can combine text, graphics, and voice—television alone can do that. The point is that the computer can switch instantly from one to the other. The receiver of the annotated spreadsheet can listen to the comments if he or she wishes, or ignore them and examine the spreadsheet itself in detail.

DOCUMENTATION

The complexity of high-technology products is so great that they require enormous amounts of reference material for maintenance and repair. Nowhere is this more obvious than in the Defense Department. The amount of paper aboard a large naval vessel is so heavy that it affects the ship's speed and handling. There are so many manuals for a modern military airplane that mechanics can't possibly take all of them along, so they have to make many trips back and forth to the hangar—or worse, they work from memory and make mistakes. There are similar problems in the private sector.

Highly regulated industries, such as utilities and those that handle hazardous materials, have a nightmarish logistical problem of keeping employees up-to-date on the information necessary to operate legally, safely, and efficiently.

These industries are prime candidates for electronic documentation.

The computer's ability to simulate can make electronic documents far more useful than their paper counterparts. Already, manuals exist that adapt themselves to the technician's experience. A novice receives detailed instructions; the expert only has to look at the outline.

Animation is a great improvement over the incomprehensible "exploded diagrams" found in many technical manuals. Likewise, the computer's ability to reproduce other sights and sounds can be used in diagnosis. "Does the engine sound like this?" "Does the semiconductor wafer have discoloration like this?"

VISUALIZATIONS

Scientists and engineers increasingly use visualization to understand how objects that are too small, too big, or nonexistent might look. Computers can model everything from atoms to the universe, yielding new insights into the workings of all sorts of things.

Further, computers can show what an unbuilt building or next year's automobile design will look like. It is becoming easier and easier to create artificial environments that are much easier to construct and change than the real thing. Models of buildings allow architects to take their clients on a tour, or show a city planning commission just how the skyline and shadows will change.

THE FUTURE

As an interactive video producer, you are at the forefront of an evolution in computing. You're also on the leading edge of a revolution in communications.

There are many mysteries about interactive multimedia. Although we can see that these tools and products will appeal to many people for many purposes, we know very little about how to create them so that they truly are useful to a wide range of people. The controls that work for cable television, VCRs, word processors, databases, and spreadsheets will only work for the evolutionary applications, not the revolution.

The rise of information-rich software is the revolution. The merger of audiovisual data with computer programming demands new human-computer interfaces, if this revolution is to grow beyond the minority that is turned on by raw technology.

You can make that happen by learning everything you can about the things that do and don't work. You can make it happen by subjecting your creating to testing by real people. Test, test, and test more—there are many dead ends still to be discovered.

The long-term promise of this new kind of communication is to create new communities. Business already has demonstrated the power of electronic mail and teleconferencing and other computer-mediated communications to help them run more efficiently, to allow large numbers of widely dispersed people to work together.

Less expensive and easier-to-learn versions of similar tools can help address enormous social problems. Good communications can reduce the need for travel. Parents can work at home part of the day, to be home when their children arrive. Educational software and links among parents and teachers can allow parents to become more involved in their childrens' education.

Your work today is setting the stage.

Nick Arnett
President
Multimedia Computing Corporation
Santa Clara, California

Preface

The steps of multimedia application development are very similar to any major project. There is a planning stage, a production stage, and a final stage where products are made ready to deliver to a client or to the public. The complexity and requirements of each of these stages is dependent on the particular project. You would not attempt to design an office complex, or take over the controls of an airplane, or even to bake a cake from scratch, unless you had some interest, knowledge, and understanding of the scope of what you were about to do. The goal of this book is to provide you with enough information about the production of a multimedia application that you can begin to develop your own multimedia products.

If you have produced media before, the information here will help you to understand what is different and unique about production for all-digital multimedia; if you have worked on software projects before, you will gain insight into the special requirements of audio, video, and still-image production. The complete novice will find this book a place to begin to learn what is required to begin, and how to put a multimedia team together.

The information here is presented in three parts that correspond to the three stages of project, or product, development—planning, production, and the final stage of releasing a product. Throughout the book, you will find production tips and techniques that have been used by other multimedia developers. The orientation of the authors is multimedia production. Each chapter is presented from the viewpoint of a producer, and is written so that a wide audience will be able to understand and apply the ideas and facts presented.

The first section of the book is dedicated to the project planning stage. In Chapter 1, you read about multimedia in general, and DVI® applications that are in use today. Chapter 2 is an introduction to the DVI® hardware and software, and the basic technical concepts of video compression and decompression.

Next, Chapter 3 provides more detail on the key production steps for developing DVI® software applications. In Chapter 4, you will find information on the people, resources, and talents you may want to tap in multimedia production. Chapter 5 provides more detail about the DVI® application development workstation, including the hardware and peripherals needed to complete the workstation.

The last two chapters of Section 1 share "tools of the trade" for project planning, and will spell out ways that you can use popular software products to help you control budgets and time lines of the project. Creative tools will also be unveiled to help you document the project with storyboards and scripts.

Section 2 of the book is a more detailed treatment of production. After a more in-depth look at the theory of digital motion video compression and decompression and image formats, each media element is covered in detail. Information on audio production, video production, still-image digitizing, and compression are all presented. Each of the ActionMedia 750 Production Tools is introduced. Tips, shortcuts, and features of the ActionMedia 750 hardware and software are covered in this section.

Particular tips in creating graphics and in integrating visual material, such as motion video, stills, and VGA graphics in the DVI® environment, are provided in Chapter 9. Chapter 10 is totally dedicated to still images as a media element in DVI® applications, while Chapter 11 is devoted to motion video and the special production techniques to keep in mind while producing video for the all-digital environment. Audio production is covered in Chapter 12. Chapter 13, the last in Section 2, is an overview of programming and authoring for DVI® technology. These chapters will help you to decide how you design a production that will result in a successful application.

Finally, Section 3 is dedicated to the final steps of getting your title ready for distribution. If you are preparing your title for CD-ROM or other magnetic media, including distribution on a Local Area Network, you will find the necessary information in Chapter 15. This section will also give you some tips about how to test your application, ways to store and distribute production among team members, resources about copyright law, and information about the future that you may want to keep in mind as you develop applications today.

In total, our goal is to provide you with a valuable resource to jump into the exciting world of multimedia production. There are many valuable lessons in these pages, and we are pleased to be able to share them with you. We hope to lead you down the path far enough that you can uncover more ideas and experiences as producers. Good luck!

Mark J. Bunzel
Sandra K. Morris

Acknowledgments

Preparing and writing a book is an ambitious undertaking, especially when you hold a full-time job. We would like to thank Bill and Deena and our families for the patience and support during the weekends and evenings this book came together. To our children, Gordon and Garrett, as you grow up we hope the information contained in this book will inspire multimedia applications you can learn from and enjoy. When this happens, our goals of this book will have been met.

There are many people at Intel who have been extremely supportive and contributed directly and indirectly to the writing of this book. We would like to thank Tom Trainor, the General Manager of Intel's Multimedia Products Operation (MMPO), located in Princeton, New Jersey, the home of the DVI® technology. Tom contributed with his support, vision, and guidance. At MMPO there are a number of people who have contributed directly and indirectly to this effort. We would like to thank Rick Stauffer and Al Korenjak for reviewing portions of the book, and Karen Andring for her excellent copy editing and suggestions. Mike Keith, Kevin O'Connell, and Rick Yeomans helped to confirm technical details about DVI® technology. Many other members of the Intel team at MMPO have also contributed to the knowledge and techniques described in this book.

The staff at Avtex Research has been an enormous help in assembling the content and developing many of the production methods described throughout the book. Special thanks go to Leslie Service, Debra Lyons, Ken Wiens, Lisa Ramirez, Derek Wade, Eric Hards, Laura Phillips, Deborah Baker, Natalie Brunello, Jeff Dods, Karen Moultrup and Sue Drescher for their help in assembling this material.

Many of the application screen designs used as examples in this book are the creative work of Nancy White. The cover design and artwork was created by Chad Little of Tracer in Phoenix, Arizona. David O'Dell and Cory Law of Intel's Corporate Graphics department created the illustrations used throughout the book.

We would also like to thank Berit Osmundsen-Wick of Intel, and

Dan Gonneau and Nancy Sileo of McGraw-Hill who encouraged us and have helped to keep the book production on schedule.

There are a number of other important contributors who have also helped us to learn more about the technology and who influenced the preparation of this book. We thank you for your contributions and look forward to your sharing with us in the success of the multimedia industry.

Mark J. Bunzel
Sandra K. Morris

Multimedia Applications Development

Section

1

Project Planning

Chapter

1

Multimedia Basics

THE MEDIA REVOLUTION

Human beings have always needed to communicate. From ancient storytellers to modern-day film, we have always searched out new ways to pass information, emotion, and insight to others. We chuckle at the quotation of Thoreau challenging the invention of the telegraph, questioning if anyone in the states of Maine and Texas would ever have anything to say to one another. Today, it seems that each new advance in technology centers around new ways to store, retrieve, and transmit information.

The collection of ways that we communicate is called *media*. Corporations use media to stay competitive. Educational institutions use it to teach and stimulate. Entertainers use it to tell stories. Training professionals and information specialists have long understood the value of media to help gain and maintain attention, as well as to deliver content. Advertisers and entertainers take advantage of media to stimulate human senses and to heighten emotion about a story or a particular product. We use media to send messages.

What is multimedia?

Multimedia has grown out of a variety of media disciplines. As early as 1978, Nicholas Negroponte, a scientist at MIT's Media Laboratory, predicted the fusion of the broadcast, print publishing, and computer industries as the direction of communications technology. Today, that vision is a reality. The personal computer (PC) can now deliver all types of media—text, still images, graphics, audio, and full-motion video. More importantly, the personal computer brings to all of this media two other important functions: the ability to present this media in an integrated way and in an interactive way.

Interactivity is not a new idea. Though it is often attributed to the computer, we interact with many forms of media. When you scan the front page of a newspaper, and then select what articles you want to read, in what order, you are interacting with it. When you tape a television program at a certain time and watch it later, you are using technology that allows you to interact with the television. Interactivity is probably most often associated with the computer, however, because the computer is so good at sorting, searching, and cataloging large amounts of information.

When you interact with the newspaper, you shuffle through large pieces of paper and sift through a lot of unwanted print material. With the computer, you can almost instantly access the information requested. It appears on the screen, in real time, cutting out any reference to the unwanted. While you can tape a television program and watch it later, you are still dependent on taping when the program is scheduled for broadcast. You cannot interrupt the broadcast to search for other related information. The computer allows you to store information so that it can be viewed at any time, interrupted, repeated, and changed or enhanced in a variety of ways. Imagine merging the television news, for example, with the newspaper, linking articles that provide the history of an issue with contemporary interviews. This is the power that the computer brings to media.

Multimedia and technology

Technology moved us toward interactive capability with video and still images when videodisc and interactive videotape technologies arrived on the scene over 15 years ago. These technologies linked analog video with digital information, such as graphics and text, by using the personal computer. The computer controls the videodisc player or tape player, so that you can jump to certain scenes or "frames" of information from the video source. These systems were useful in some corporate training settings, educational institutions, and in retail stores.

A technology exists today, DVI® technology, that allows all of the information, including full-motion video, still images, audio, text, and graphics, to be stored and distributed on an all-digital source. The personal computer can manage all of the information as digital files, and can deliver it to the computer user from CD-ROM (*Compact Disc*

Read Only Memory) or hard disk. All of the information can be sent over networks or telephone lines. This capability turns the personal computer into a multimedia tool. Now, the interactive capability of the computer can be applied to all types of media.

In the following pages, we will explore many of the uses, or applications, of this new capability. Technology and industry gurus, business leaders in hardware manufacturing and software development, agree on one thing—the future of this technology and its potential lies in the hands of application developers. It is application developers who will take this capability and build new, creative ways to educate, inform, and entertain.

WHAT IS A MULTIMEDIA APPLICATION?

Multimedia means that software can now be the electronic equivalent of an audio instructional tape, a print catalog complete with sounds and images, a travel brochure with moving pictures and an on-line reservation system, or a business presentation complete with pictures, animation, video, and narration. A multimedia application can be as simple as augmenting a real estate database with pictures, to adding an audio help function in one or two languages, to a complete training simulator of a dangerous or expensive flight mission.

Obviously, this diversity calls for a range of skills and information about production, interactivity, and the subject matter that will be explored. Multimedia can be compared to desktop publishing. The cost of creating an application and the skills needed to complete the work are dependent on the expectation you have for the outcome. If you want an internal newsletter that quickly communicates a few key ideas, you will probably create the text and graphics yourself, print it up, and mail it out of your own office. On the other hand, if your goal is to distribute a newsletter to the customer base of a multimillion dollar company, you will probably enlist a team of experts that will support you with content, layout, photography, design, and printing.

The same can be said for multimedia. DVI® technology was developed to enable hardware manufacturers to build multimedia personal computers. In turn, end users are able to create everything from a business presentation to complex training and information packages. This new media requires new skills for software developers, but they do not have to be learned all at once.

Content is the new element

The difference between creating software for the personal computer today and creating multimedia applications is *content*. Applications for personal computers, like spreadsheets and word processors, rely on content that is provided by the user. A multimedia application is more like a movie or book. The creator decides on some content—a business training problem that needs to be solved, a sales presentation with interactivity built in, or an arcade adventure game with characters and scenes from a popular television series. Content must be created and produced. Then, software is written to manage the content. Software *and* content comprise a multimedia application.

Content of a multimedia title can be just about anything from a series of still images arranged to simulate surrogate travel through distant lands to short audio clips to describe functions in a word processor. The development process can be as simple as adding pictures to a database to creating an environment where multiple users on a network can add video messages to corporate reports or memos. Similar to a book or movie, you can create anything from an epic to a cartoon short, and all the shades in between. The type of application you will create depends on your goals, market, and resources, much like the producer or editor of any print or video material.

MULTIMEDIA SOFTWARE DEVELOPMENT

There are many similarities between how a book or movie is produced and the development of a multimedia software application. Unlike a movie or a book, however, you can create much of what you need for multimedia software right at your desktop. A personal computer, equipped with the boards and software for DVI® application development, opens the world of multimedia software development. This new power requires that the software developer be knowledgeable of a broad range of production.

Audio production and still photography are two examples of skills that may be needed. If motion video is going to be added to software, or if a brand-new title is going to be created, the skills needed will include content expertise, video production expertise, and visual art talent. Technical talent is always needed, since a multimedia application is ultimately a piece of computer software. But software programming is not the central activity, as it may be in the creation of database software or a word processor.

The talents of production

There are three skill areas that will be needed to create a multimedia application:

1. Interactive and content design
2. Graphic and production design
3. Software design (or architecture) and programming

This does not mean that each multimedia application (or title) will require an expert from each one of these areas. Like newsletter production, there will be many cases when people's talents will overlap.

If, for example, you are adding an audio-video help screen to a spreadsheet, you will probably need to have someone working on the team who has graphic and still-image production experience, audio production experience, and experience with how the help function in the spreadsheet works. You will also need to have someone on the team who is considered an expert on the types of help that will be most useful in the product, and who can design what the interactions will be like for the user.

On the other hand, if you are adding still images and some motion video shot at a corporate picnic to a presentation to the human-relations department of your company, you will probably be able to do this alone, or with the help of someone who knows how to use the presentation software.

You may have noticed that we refer to most of the talent needed to create a multimedia title as *design* skills. Much of the content that will be produced for multimedia applications is visual, so it is natural that we borrow terminology from the disciplines that are visual in nature.

But design of a multimedia application goes further than deciding what colors and images appear on the screen. In a nutshell, there is more to multimedia design than meets the eye. The look of the application is important, but of equal importance is the "feel." How does it flow? How easy or difficult is it to use? What are the interactions that are encouraged or discouraged by the screen and by the way the program acts?

Interactive multimedia design

In any type of personal computer application development, programmers often speak of a top-down versus a bottom-up development process. In bottom-up programming, the programmer begins with a blank slate, and writes the code to make each piece of the application work. As the pieces are built, they are then tested and integrated as a whole. This process has its advantages—a programmer can create software that fits any particular operating system and build onto it in manageable pieces. As the layers are built, expectations can be raised or lowered accordingly.

In a top-down approach, a design is developed first, and a skeleton prototype of the user interface and how it may operate is implemented. The programmer peels these layers back and addresses the interaction of software, hardware, and the design of the application at each level until the design is implemented.

Media, like film and television programs (except live broadcasts), takes a top-down approach. The story is written and refined. Then the actors are identified to play each of the characters. Then the production is planned. The filming takes place; audio and other visuals are collected. Last, the elements are put together so that they reflect the story that was originally written. Text materials are usually put together this way as well. The goals and objectives are solidified, material is written, visuals obtained. These elements are put together in a layout, and finally, the piece is printed, collated, or bound.

Like its cousins the film and the book, development of the multimedia title is a top-down process. The phases of development begin with a design phase, and move on to a production stage, when audio and video are collected. Programming can go on during the production stage, and continues on to the product testing and documentation stage, when the final application is made ready for delivery to the client, or the publishers (see Fig. 1.1).

The first questions in the design phase are obvious: What information are you trying to convey? Is multimedia the best way to convey it? A second set of questions is more difficult: What do you want the experience to be like for the user of your title? What are the most important messages?

The design process is the most important step in development. During this process, you set goals that will determine how the application will work, and what content you must gather. You will begin to gather all the information you need to continue into the project planning phase, where you will have to tackle hard questions about resources and budget.

Defining a title's goals–a framework

One way to define your multimedia application is to outline three important factors that will effect your design:

1. Subject
2. Setting
3. Audience

Defining content

The subject of an application can determine many of the variables that need to be considered during its design. The subject of the application determines the content, and to some extent, the purpose of the title. The first questions to be answered are:

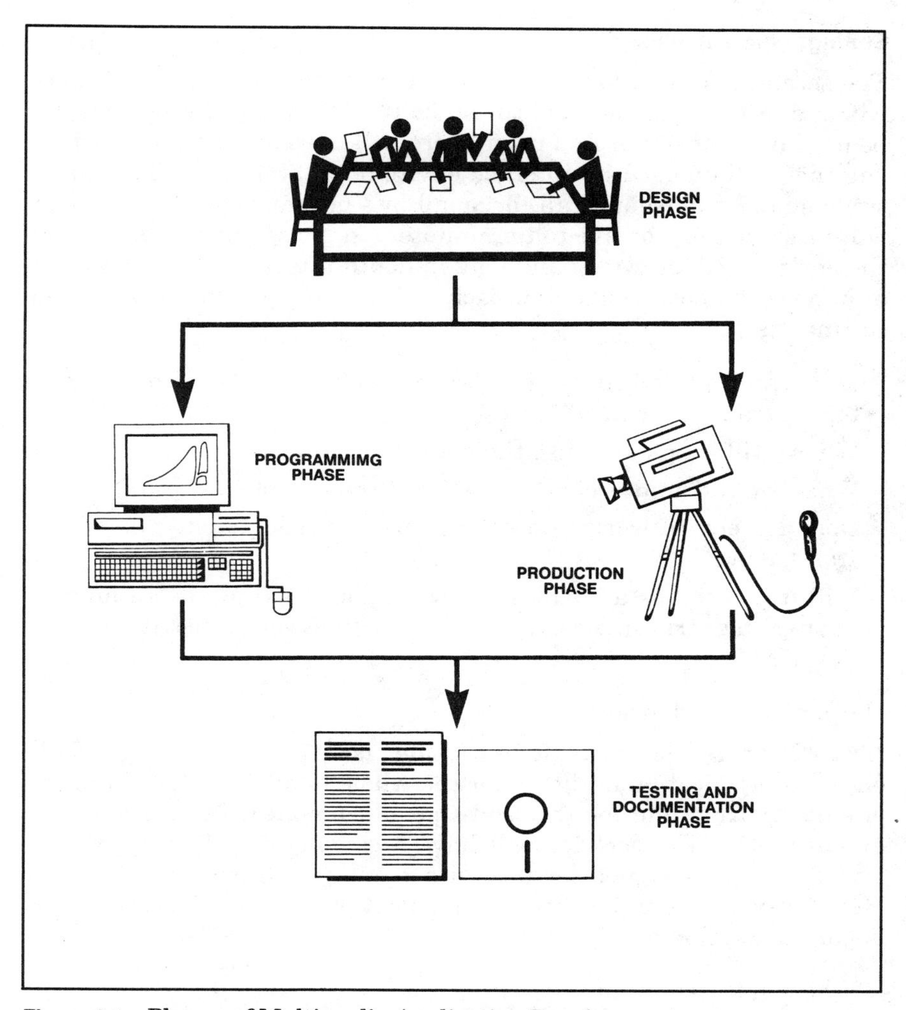

Figure 1.1 Phases of Multimedia Application Development.

Is the subject technical? Entertaining? Instructional? Sales-oriented?

What are the problems involved in conveying the information currently?

How will multimedia solve them?

What parts of the material are most visual?

How will audio augment the material?

Is random access, linking of topics, and branching important?

Setting: Where it is used

The second topic to address in the design phase is the setting. Where, when, and how will the application be used? An application that will be used by a consumer in a retail store will have a different tone from one that will be used in an office environment. Some settings, like a trade show or museum, call for small bits of information that a user will interact with for five to ten minutes. In a classroom setting or on the desktop PC, browsing and long in-depth interactions with material may be the best design approach. A few questions to ask about the setting are:

Is the information time-critical or not? Will it be used on the desktop, or trade-show floor?

Where will people best use the material?

What are the unique characteristics of the setting?

Does it need to work into or support a larger display, or does it stand alone?

Will it be used as a business presentation, for desktop training, or in on-going performance support in a business or industrial setting?

The audience: Your market

A third general design topic to consider is: Who is your target audience? An application designed for the whole family will have a different flavor from one for thirteen-year-old females, for example. An executive training package will have a completely different orientation from one designed for use with new hires fresh out of college. Here are some sample discussion questions for your team as you begin the development process:

Who is a typical user?

What is her or his level of experience with computers and with the subject matter?

Is literacy a concern?

What does this group respond to?

What are the client's expectations for this group?

A good discussion around a specific problem or business opportunity will bring out additional questions, and force you to document and design a specific application to solve a problem or meet a business goal. The important thing to remember in this stage of the development process is to know as much as you possibly can about the setting, audience, and the subject of your proposed title.

During your initial discussions, you may want to bring in experts in the content area you are discussing. If you are thinking about a point-of-sale application in a grocery store, you will want to talk with manufacturers, grocery store managers, and maybe even consumers. If you are interested in creating an educational application, you will benefit from talking with teachers and publishers of educational material.

Specific content will come later. This is not the time to unearth the minute details about the content you will produce. It is the time to determine if multimedia is the right direction to take, and to understand what type of content production will be necessary to meet the needs of your audience and market. The analysis of the business problem or instructional problem you are solving is the key to determining if multimedia software is the solution.

Visit the sites where you propose to sell the application, talk to typical end users, and understand the culture of the organizations that will use this software. This understanding up-front will pay off at every level of design and title development.

Options in design: Multimedia element

Another important set of considerations to examine during this phase is the type of media you will be able to deliver. In later chapters, each of the media options will be discussed in more detail. The range of options available to you with DVI® technology is comprehensive, including:

Still photography

Motion video—RTV (*Real-Time Video*) and PLV (*Production-Level Video*)

Graphics

Animation

Text

Audio—music, speech, sounds, etc.

Interactive program to "drive the presentation"

Each of these options can be presented in a variety of ways:

Stills:

Full frame

Partial frame

Differing resolutions and pixel depths

Still transitions (special effects)

Animation:

Computer-generated graphics

Video animation

Full-Motion Video:

Production-Level Video (PLV)

Real-Time Video (RTV)

Full or partial frame rate

Full or partial frame size

Audio:

Full range of quality levels from AM to FM stereo

Computer Graphics:

DVI graphics

VGA graphics

Interactive Program:

Implemented in C (programming language), or an authoring tool

This list gives you some idea of the types of decisions and options you will have in your application. More detail about the production of all these elements is the subject of Section 2.

MULTIMEDIA APPLICATIONS TODAY

There is no telling what future applications will be developed for multimedia. Today, applications are being used in corporate training, manufacturing training, education, sales, and entertainment.

One of the most immediate results of adding multimedia capability to the PC is that all end users will benefit from a friendlier, more human interface. Imagine the spreadsheet or business presentation software with an audio and video help feature. When you are confused about a function, or how it is used, a human tutor appears to show you examples of how to use it. Tutorials would be included to help first-time users of these packages.

While these applications seem apparent, they have not been the earliest developed. Most of the productivity software on the market today does have some level of text help built in. The most immediate need for multimedia is in areas where alternative solutions don't yet exist, or where a text and graphic computer-based solution is not viable.

Figure 1.2 Florida Traveler Information System.

Shopping by multimedia

Interactive multimedia kiosks located in information sites around Florida, as seen in Fig. 1.2, can be used by visitors to learn about the local sites, flora and fauna, and recreational facilities as they travel. Information kiosks like this one are currently in use in Florida, and in airports in France.

These systems provide an inspiration for a variety of consumer products as well. Kiosks can be placed just about anywhere, from furniture stores, to bookstores, to grocery stores. CableShare, Inc. in London, Ontario is using DVI® multimedia to send images to homes for a home shopping application. Real estate shopping from home, accessed through the touch tone telephone, is one product offering, illustrated in Fig. 1.3 (see color plate).

Multimedia in manufacturing

One example of the use of multimedia is in Intel Corporation's manufacturing area. As a leading producer of semiconductors and microprocessors, Intel faces the task of training a large, high-technology, manufacturing work force. The manufacture of Intel's products requires a variety of precise skill sets, including the visual inspection of parts during the manufacturing process. This task is very visual. A wafer, the basis of semiconductor manufacturing, is built in layers.

Each layer must be inspected under a microscope in order to avoid defects in manufacturing.

To complicate the training process further, paper cannot be introduced to the manufacturing environment because fine particles can contaminate the factory. This causes a real business dilemma for the training managers: How can you adequately train inspection operators for this complex task without reference manuals?

The answer is a training and job-aid multimedia application for the inspection process. This application uses still images taken directly from the microscope, video from the fabrication floor (factory), and instructional audio and graphics used in the initial training of inspectors. The application software is provided for training and in the inspection area for operators to access as an electronic reference while on the job. Instructors use the software in classrooms, and it is maintained in the library training area of the factory. Systems with the application are also available in the inspection areas of the factory.

If a certain questionable image appears in the microscope, the operator accesses the multimedia database in order to confirm whether there is a defect present, and obtains instruction about how to proceed. If an inspector loses sight of exactly what is to be examined in a particular layer of a wafer, the database is examined, and training accessed immediately. A sample screen from this application is shown in Fig. 1.4 (see color plate).

The benefits of this application are:

A paperless training environment

A decrease in training time and increase in retention

A method of job support that increases the efficiency and accuracy of any particular inspector

The results are impressive. This multimedia application, and others like it, have the potential to tutor workers while on the job. Multimedia increases access to information that was once only available through experienced coworkers in combination with reference manuals and text.

Corporate training

Interactive video training has been used in corporate training, with much success, for some time. Many studies exist that show that interactive training increases retention, decreases costs, and lessens the amount of time needed in training sessions. All of this can translate into well-designed DVI® applications. In addition, the added benefits of digital distribution, and the lower cost of using existing personal computer platforms for delivery, make this training more accessible and more affordable.

New York Life insurance company is using DVI® technology to show off their Boston office to new recruits. Interviewees are invited to "meet" the general manager and other employees, and to tour interesting sites of Boston while they are visiting the office.

Southwest Research Institute in San Antonio, Texas, has developed a number of DVI software applications, including a system that was delivered with a paint-stripping robot. The robot was also designed by SWRI. It is the size of an airplane hangar, and it is used to strip paint from airplanes. The interactive training system that comes with the robot uses DVI® technology to deliver full-motion video, graphics, audio, and text in an interactive training application. This application, shown in Fig. 1.5 (see color plate), teaches operators how to use the robot, and how to teach the robot new movements.

Education

One of the first DVI® applications ever demonstrated was Palenque, a tour through a Mayan ruin controlled by a joystick. Children and adults who used this application were able to "walk" around Palenque, and could access information from characters like C.T. Granville, a young companion, along with an archaeologist and a botanist. Information about plants and animals, history of the Maya, and maps of the site were also available. Today, Palenque is being completed by Bank Street College of Education to be placed in museums around North America.

Another group in Canada, the Lindsay Library, in conjunction with Pixel Productions, is using DVI® technology to document the town's history (see Fig. 1.6a—color plate). Members of the library staff are collecting historical photographs, video and audio, and producing new media to include in a historical database of the town. This application is drawing on the experiences and the memories of town figures, creating a town masterpiece that will preserve the story of Lindsay for all of its inhabitants. DVI® technology allows all of the town members to participate in the collection process.

Transmission of images—VideoFAX

A group of physicians in Tennessee and Florida are using DVI® technology to transmit video information needed for diagnostics. Today, ultrasound technology is used to examine the heartbeat of a fetus almost routinely. These tests are typically recorded to videotape by an ultrasound technician, and then read by a radiologist.

The ultrasound data in this application is also recorded, on hard disk, instead of videotape, as shown in Fig. 1.6b (see color plate). The data is then transmitted over standard telephone lines from Florida to Tennessee, where it is reviewed by a specialist. All of this occurs in one hour, the typical appointment length for a pregnant patient. Since most ultrasound centers do not have a specialist on staff, this eliminates sending patients to hospitals far away from home for this procedure.

Opportunities in multimedia

The problems that are being addressed by multimedia titles have one thing in common—variety. Multimedia titles can be found in all industry market segments, including medical, government, higher education, manufacturing, school-age education, point-of-sale for a wide range of consumer products, and entertainment.

All of the ideas and applications we have seen are exciting! Yet, they are only the tip of the business that is being created as a result of all-digital multimedia. It is difficult to imagine, and rather humbling to realize, that in 10 years we will look back at these applications with nostalgia and a touch of affection.

Today's multimedia software programs have been compared to early television shows that mimicked radio, or early training videos that consisted of an instructor speaking in front of a blackboard. We are much smarter about how to use media, and about how to interest, motivate, and teach people. All of what we already know can be brought to bear on each title. History repeats itself, no doubt, with our new multimedia titles; we all have a lot to learn. Yet the potential of genius is in each new title, just as it was in early films, radio, television, and software applications.

THE KEY INGREDIENT—CREATIVITY

The hardware and software systems are certainly a necessary and very important part of this revolution in communicating. They are full of intrigue, and understanding them is essential to good design and to good business. Any person who is part of a multimedia development team should learn as much as possible about how the systems work, for one reason—so that the best titles can be designed. In the remaining chapters of this book, you will find a wealth of information that will empower each of you with technical tips and techniques that will move you forward in your work.

But knowledge of the system is not enough. Our last message here has nothing to do with the technical process, but with the creative one. Multimedia is the revolution that will change the personal computer. It also has the potential to change the way we work and the way we play. To do that, the developers of titles must make sure that they allow all the creative energies possible to flow into the design of titles.

The "keeper of creativity" for each team—there may be several—must be committed to the fundamentals of an open, creative environment. Following a few basic tenets of brainstorming will help ideas bounce around the frame of your first multimedia concepts. Let creativity push on its walls to include more and different paths.

Use metaphors for thinking about problems. Humans use metaphors to think and to create, both in language and in the analytical subjects. Metaphors allow people to think of ways to build on ideas and to use experiences.

Give the team a safe place to consider the outrageous, the unthinkable, the undoable. And encourage it. Try to turn things "upside down." When someone talks about flight, for example, someone else may pick up the theme of the subway or undersea exploration. Explore it. You never know . . . it may work.

Keep ideas alive as long as possible. There are many shades of gray—in fact there are 16 million colors available on a DVI® system. Any idea may surface as the right one in the course of a project, or in the course of a team's experience with many projects.

Push for excellence. Push your team and the technology, and commit yourself to creating "no-compromise" titles that we can all look at in amazement and with pride.

However the talent comes together to create the multimedia revolution, the common goal is to design, create, and produce an integrated work on a personal computer. This requires that someone, whether it is the project manager or programmer, learn and implement techniques that are unique to the new, multimedia delivery platform. Engineer, artist, and programmer must learn how the addition of media content affects what they do.

This book is a talent-building resource. For the fully complemented team, or for the individual title producer, whether programmer or graphic artist, the information in these pages bridges the gaps of knowledge and expertise required to create multimedia titles. We do not expect to eliminate the learning curve that each professional will have as he or she embarks on title development. It is our goal to help each reader understand the process and avoid costly mistakes. We hope to unleash your creativity and productivity for the production of meaningful and exciting content.

Chapter

2

DVI® Technology— An Overview for Developers

As someone who is interested in the development of multimedia applications, you are probably familiar with the computer as a tool. We often take for granted how technically sophisticated our culture is as a whole. Consider the modern home, for example. The compact disc player, dishwasher, microwave oven, VCR, and TV all have controls that would be complicated to someone who had never seen them before. The computer is usually considered to be out of the range of mainstream appliances, and you are still considered to be special in some way if you can operate one with ease.

Of course, the introduction of the *Graphical User Interface* (GUI), and other easy-to-use software, has made this less true. Yet, the computer has not reached the level of the microwave oven in the ease-of-use category. The addition of multimedia applications to the personal computer may change that, but first the developers of these applications must understand the systems that will allow them to build the "microwave" of PC's.

Just how much do you need to know about how DVI® technology works in order to create an application? That question is a very difficult one. Today, you probably need to know more than you will five years, or even two years, from now. As the technology that enables multimedia becomes more transparent to developers and end users, and as more applications are created, the specific knowledge of how the technology works will become less important.

This chapter will cover some basic information about DVI® technology. As you read, please keep in mind that these concepts will come up again and again throughout the book. If you do not understand something when you first read it, make a note. We will probably answer your question in some other context later.

DVI® TECHNOLOGY BASICS

It is certainly not necessary for each team member on an application production to know everything about the technology. Some will need to know more than others, or different things from others. For example, a video producer may want to know how compression will effect the motion picture image, while a graphics designer will want to know what graphic formats will be acceptable for the software.

As in any digital system, the number of interactions by the user, and the type of graphics or images using in a software application, all require system memory and processor (CPU) performance. An interactive designer must know how quickly certain interactions can take place, or what type of graphics can be displayed with motion video, and other facts about memory and performance.

Most likely, you will find out about features of DVI® technology in the same way that you find out about features of your word processor. You have to understand the functions and features of a word processor to begin using one, but the specific implementations, like setting tabs, creating charts, and importing graphics, are things that you look up and use as needed.

A good rule of thumb as you take up the multimedia palette with DVI® technology is: The more you know, the better. We have selected the important concepts about DVI® that we think will help you begin. For a more technical discussion of these ideas, you may want to consult *Digital Video in the PC Environment* by Arch C. Luther. This book contains a discussion of the theory behind digital video compression, and goes into detail about technical aspects of DVI® technology that will not be covered in detail in this book.

The key concepts we will cover in this technology overview are:

- Video compression algorithms
- The i750 video processor that makes video compression and decompression possible
- System software, tools, and authoring systems that can be used to create interactive, multimedia applications

WHAT IS DVI® TECHNOLOGY?

At the most fundamental level, DVI® technology is a set of video processors and software that give manufacturers the ability to create a digital, multimedia personal computer or platform. These video processors, the i750 PB/DB (pixel processor and display processor), are high-speed, special-purpose computer chips that are dedicated to the task of compressing, decompressing, and displaying video in the personal computer. Manufacturers use these processors to build add-in multimedia boards, or to build multimedia capability right onto the motherboard of a personal computer. The resulting multimedia platforms are, in turn, available to application developers.

Multimedia platforms based on DVI® technology are unique because of the approach taken with compression of motion video, still images, and audio. Most multimedia platforms, including IBM's AVC (*Audio Video Connection*) and Microsoft's Level 1 Multimedia Extensions to Windows or the *Multimedia PC* (MPC), give you the ability to show digital still images, and to play digital audio on the personal computer. DVI® technology adds compressed motion video, compressed audio, and compressed stills to this array of media choices.

Compression opens up a wide range of choices to the developer. Since the media files can be compressed, they do not require the large amount of storage required for audio and still images. A second of motion video can require 22 megabytes of memory for storage. With compression, that number can be reduced to an average of five kilobytes per frame for a total of 150 kilobytes per second. Because of the large file sizes required for multimedia applications, compression is highly desired.

Without this compression, motion video storage and playback on a personal computer is just not possible at normal data rates. Motion video is very data intensive. A single 30-second clip of uncompressed motion video playing at 30 frames per second takes over 500,000,000 bytes (500 megabytes) of storage. DVI® compression software, working in concert with the Intel i750 video processors, allows over an hour of motion video to fit into the same size file.

Compression is also a key factor in allowing DVI® applications to work over a *Local Area Network* (LAN). In order for motion video, still images, and audio to be sent over a LAN, the information must be in a digital format. In order for the application to work in a cost-effective way, the data must be compressed. Otherwise, the rate of delivery will be far too slow, and the file sizes too big to be economical.

Software applications for DVI® technology

In Fig. 2.1, DVI® technology is compared to a bicycle. The i750 video processors, the hardware that enables video compression and decom-

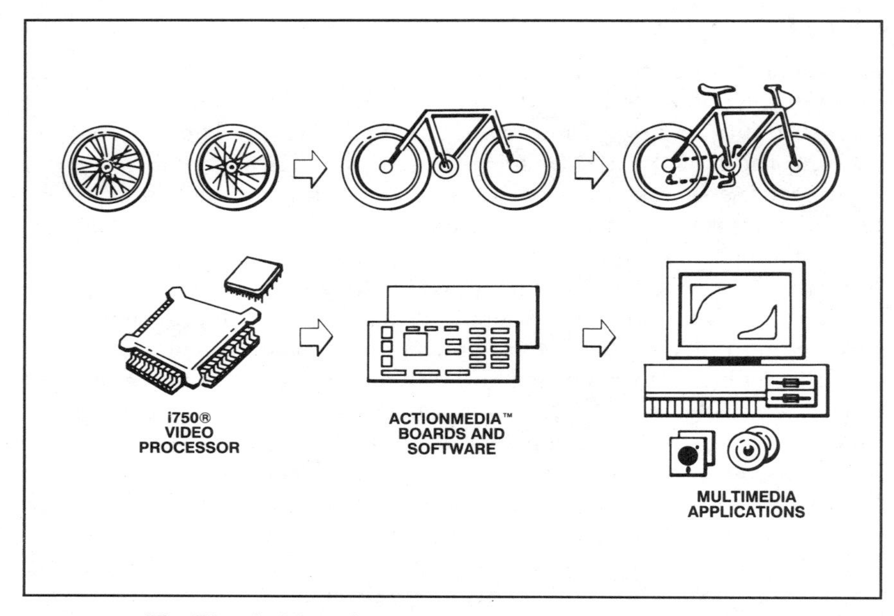

Figure 2.1 The Bicycle Metaphor.

pression, can be compared to the wheels. Software for compression is like the gears or pedals of the bike. The frame of the bicycle can be compared to the board, system, and additional software technology that makes DVI® useful to manufacturers, and ultimately the end user. Once the bicycle has a frame, a business can be built around building bikes, and distributing them. But the best part of a bicycle to the user is the ride.

The software applications that are built using DVI® technology are like the ride the bicycle rider takes. The ride can be through the mountains, around a lake, to the grocery store; the possibilities are endless. Like the bicycle ride, the software application for DVI® technology can be functional, or for sport, depending on how it is designed and developed.

Motion video compression

The i750 video processor is the key to the processing and playback of motion video that is compressed by a factor of over 100 to 1. As mentioned above, compression allows you to store and play more than one hour of full-screen, full-motion video from a standard CD-ROM. The motion video we are used to, such as videotape, television, and videodisc, plays 30 frames each second. Film plays at 24 *frames per*

WHY ALL-DIGITAL?

There have been several historical stepping stones in the development of interactive media. Most multimedia products to date rely on an analog video and audio signal being merged with a digital signal for display and playback on the personal computer. These configurations allow video from a videodisc, for example, to be displayed on a computer screen along with computer-generated graphics. The computer can be programmed to seek randomly any point on the videodisc to display motion video or stills. This set-up requires that all video, and usually audio, is stored on an analog source. The video and audio cannot be saved as a computer file. It cannot be treated like a digital file, and its distribution is limited to the physical analog videodisc or traditional high bandwidth television transmission.

DVI® allows all media—the video, stills, and audio—to be stored digitally and to be treated the same way as other digital files. Motion video, stills, and audio can be captured by the computer and stored on hard disk as files. This information can be edited; video can be cropped so that unnecessary parts of the images are eliminated; digital video files can be transformed so that they are brighter and darker; alternative audio tracks can be added to video; special effects can be written in software—zooms, wipes, fades, starbursts can all appear without expensive postproduction; graphics or text can be added to video or to still images.

But perhaps the most significant result of all is this: DVI® provides a method to compress the size of the data needed to display motion video. Motion video can be stored on hard disk or CD-ROM, transmitted over networks and over telephone lines. This capability is the key to proliferating applications and multimedia information in business and home computing. A personal computer with DVI® becomes a multimedia communications tool, and communicators and their audiences both benefit from the flexibility of an all-digital medium.

second (fps). These are the rates our eyes are used to, and accept as a moving image.

A video image converted into digital format at a resolution of 512 × 480 represents 780,000 bytes (780 kilobytes) of data. CD-ROM stores about 650,000,000 bytes (650 megabytes) of data, which means that only 30 seconds of motion video could be stored on CD-ROM. The CD-ROM is an economical, high-capacity storage media, but it delivers only 150,000 bytes of data per second. If you stored digitized, uncompressed motion video on the CD-ROM, the 30 seconds of video would take more than an hour to play back, as each frame would be constructed at 150,000 bytes of information or data per second. Motion video would look like it was being slowly painted on the screen. Figure 2.2 summarizes the need for motion video compression.

Intel's PLV compression algorithm permits over one hour of motion video and audio to be stored on CD-ROM, and matches the rate of

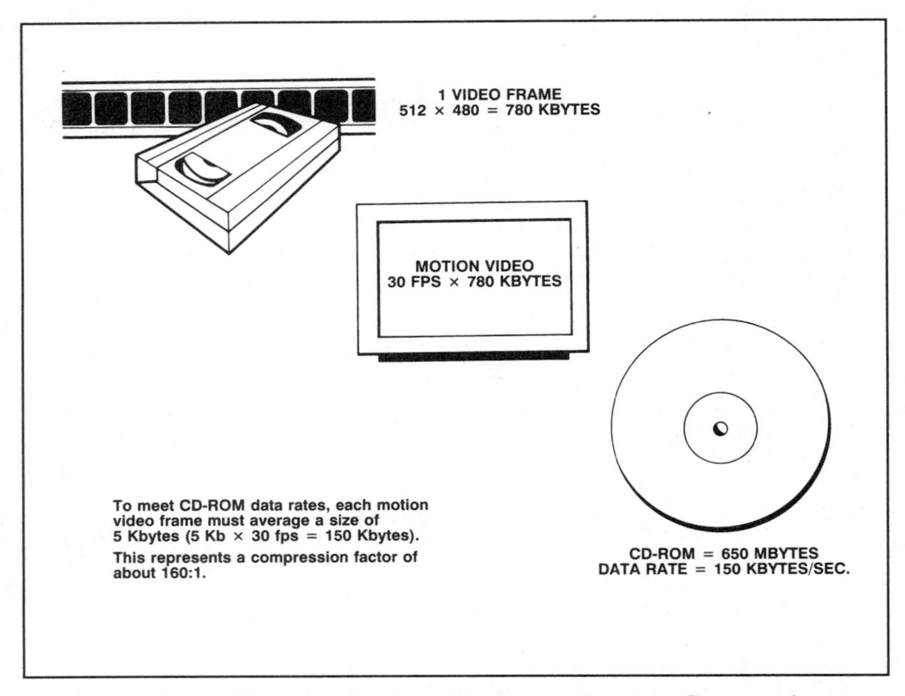

Figure 2.2 Motion Video on a Personal Computer Requires Compression.

data delivery from CD-ROM exactly. The resulting one hour of compressed, digital video plays back at 30 fps, the standard video rate, and it plays back in real time. That is, the compressed, digital motion video that was an hour long takes an hour to play back. These algorithms can also be applied to video for storage on hard disk or transmission over a network.

Video compression

One of the obvious questions to ask about motion video compression is how it is done, and how it fits into the development process. There are many choices that you will make about compression of the media that you choose for your application, and we will cover these choices in more detail in subsequent chapters. Still image and audio compression are completed by the developer at the development workstation, and the production choices available for stills and audio are issues of resolution or quality of the image or sound.

Likewise, compression of motion video introduces a set of choices that centers around image resolution and quality. Compression of motion video can be done in two ways. Motion video can be com-

WHY CD-ROM?

CD-ROM has been one target medium for multimedia titles for a number of reasons. First, CD-ROM is a significantly larger storage medium than traditional computer media, like the floppy disk or a hard disk. CD-ROMs hold 650,000,000 bytes of information, as compared to a typical 40,000 to 80,000 bytes of a typical PC user's hard disk. In addition, CD-ROM represents an inexpensive way to distribute a large amount of information. In quantity, CD-ROM discs can be replicated for under $2.00 each.

CD-ROM players are also relatively inexpensive, considering their storage capacity, costing less than one-third of a comparable hard disk. Motion video and high-quality still images are both very data-rich. The audio industry has already provided the facilities to manufacture compact discs in large quantities. The multimedia industry can take advantage of this.

Some software companies, such as Microsoft and Lotus, have already started to distribute other software on CD-ROM. Personal computer users have started to outfit their PCs with CD-ROM players. All of these trends, along with the need for an inexpensive, mass storage media for multimedia, have contributed to the selection of CD-ROM as a viable way to distribute multimedia applications.

A CD-ROM is created at a disc pressing facility, such as 3M or Discovery Systems. The creation of the CD-ROM is very straightforward. Digital data, prepared on hard disk, is typically formatted for CD-ROM using a data layout tool that makes your data conform to a CD-ROM industry standard, the ISO 9660 format. This data is then stored to digital tape. (If you prefer, most disc pressing houses have the capability to format your data for you.) The resulting digital tape is used to manufacture a CD-ROM, which holds the information. Chapter 15, "Distribution of Your Multimedia Application," provides more information about this process.

pressed on the development workstation to create motion video files known as *Real-Time Video* (RTV), or it can be compressed off-line to create files known as *Production-Level Video* (PLV).

Real-time video

Compression of motion video can be done in real time using the DVI® hardware and software products that developers use to create applications—a 386-based computer equipped with the ActionMedia hardware and software. (See box on page 31 for a listing of products based on DVI® technology.) This type of compression is called RTV for Real-Time Video, and while it can play back at a full 30 frames per second, the quality of the resulting images is lower than PLV compression, where more computing power and more time is spent on each frame of video.

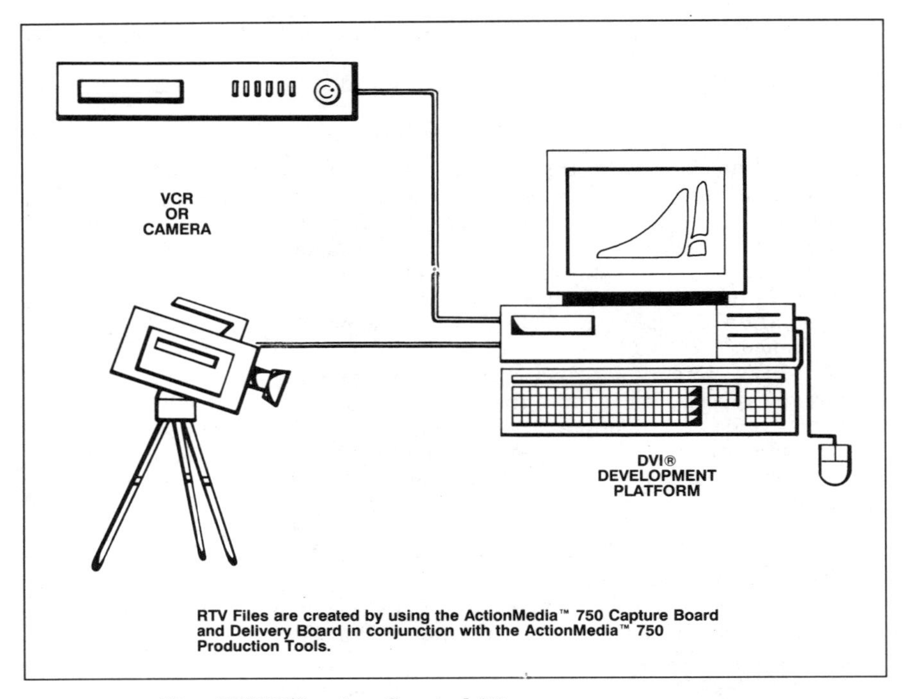

Figure 2.3a How RTV Files Are Created (1).

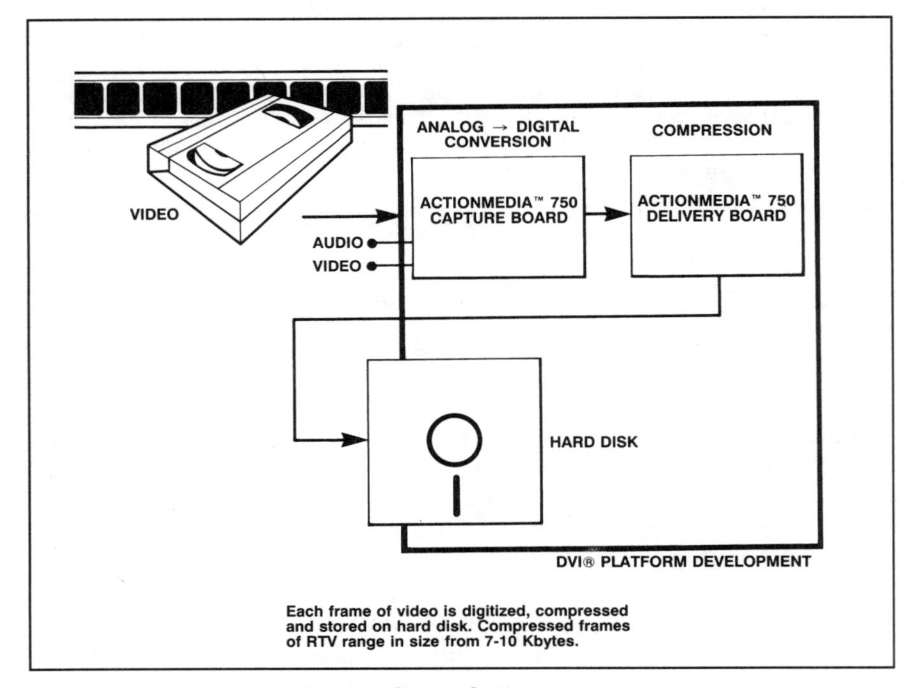

Figure 2.3b How RTV Files Are Created (2).

Developers are using RTV because there is no additional cost for the off-line compression process, and it meets the quality requirements for their application. Others choose RTV because it can be done in-house, in real time, and can be placed in the application or transmitted immediately. The VideoFAX application discussed in Chap. 1 is an example of this. Figures 2.3a and 2.3b are simple illustrations of how RTV is collected and stored on a PC.

The algorithm used to compress RTV files is different from the one used for PLV. Each image in an RTV file is a still image. RTV files are typically larger than PLV files. Each frame averages 7 to 10 kilobytes. The size of the frame, and the speed of the playback, are completely variable. RTV can be captured at 10 fps or 30 fps, for example. These files can be shown at ¼ screen, ½ screen, or full screen. The resolution of the RTV image is 128 × 120. Line doubling, or interpolation, is used to display the image at 256 × 240.

Production-level video

The process used to create PLV files is currently done at a DVI® Compression Service facility. These facilities require that developers submit a 1″ NTSC broadcast-quality tape and a special form (similar to an edit decision list) that identifies the video sequences selected for compression. When complete, the Compression Service facility returns a digital streamer tape that contains motion video files in a DOS format (see Fig. 2.4). These files can then be copied onto the development workstation's hard disk, and integrated into the application. Like the RTV files, they are played back directly through the ActionMedia delivery boards.

The PLV compression process occurs in several steps. Understanding this process can be helpful in producing optimal quality video. Briefly, the video is first turned from an analog to a digital signal. Then, the digital files are compressed by a ratio of over 100:1. The compression algorithm uses a delta encoding scheme, storing differences between frames and eliminating information that is repeated in each frame.

The compression process is completely automated. The algorithm "looks at" each frame of the video, and determines how alike or different it is from the previous frame. The first frame of a sequence is stored entirely, and is called the *reference frame*. From the reference frame forward, the algorithm stores only the differences. When the differences become very large, a new reference frame is inserted. You can also elect to have reference frames inserted by the algorithm every so often if you like. The production issues related to this are discussed more in Chap. 11, "Producing Full-Motion Video for DVI® Applications."

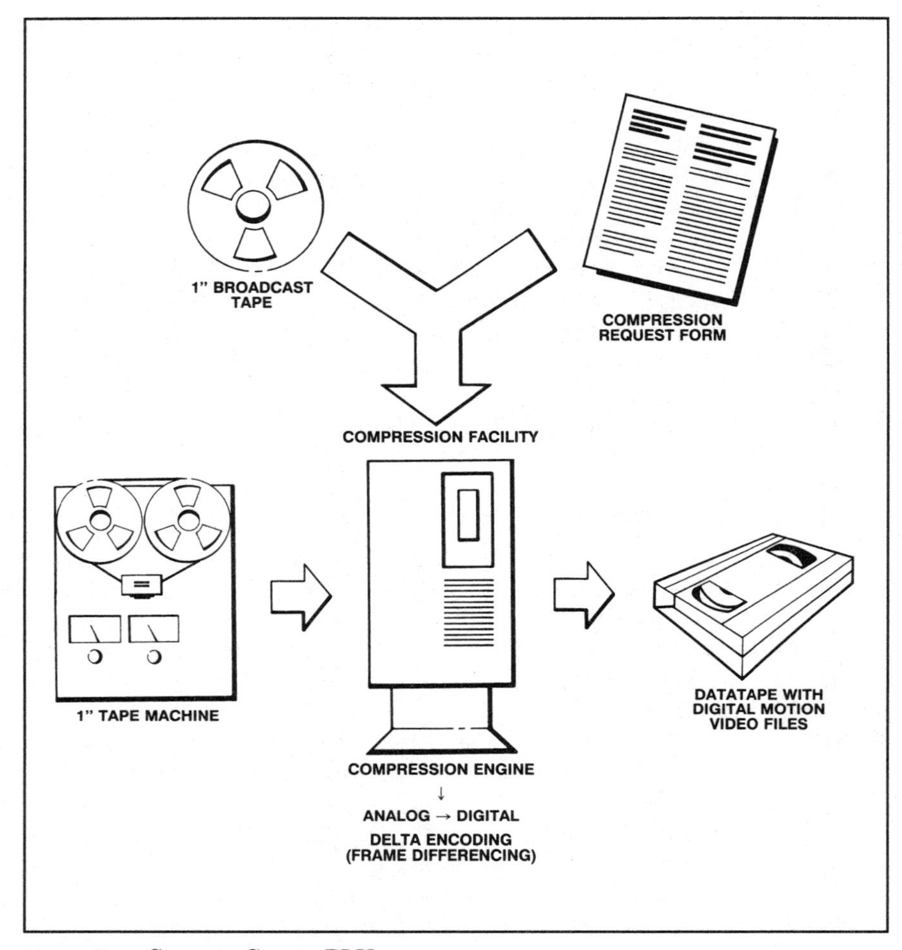

Figure 2.4 Steps to Create PLV.

Each frame of compressed motion video averages five kilobytes. This video, played at 30 frames per second, uses 150 kilobytes of data per second—the exact speed of delivery from the CD-ROM. These files are smaller than RTV files, then, and the images are better quality, since more processing power and time is applied to each frame for compression.

Compressed motion video resolution

The resolution of PLV images is 256 × 240 interpolated to 512 × 480 to match the scan rate of a VGA monitor. RTV files are interpolated to display at 256 × 240. It is important to understand, however, that the spatial resolution (the number of horizontal and vertical pixels or "dots") of a motion video file does not relate directly to quality as it does in a graphics file. Because the image is changing rapidly, and because it is rich in color, the quality of compressed motion video is much higher than the pure numbers would suggest. We have found

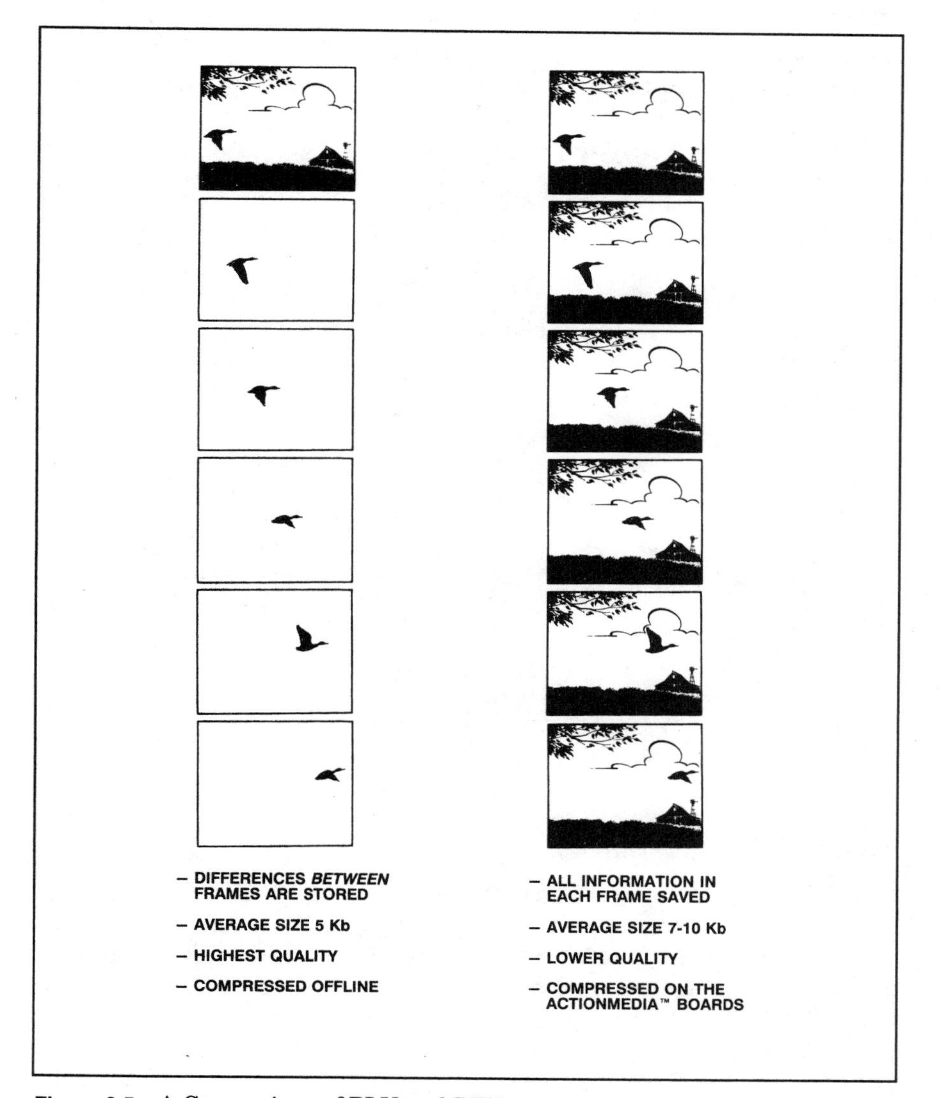

Figure 2.5 A Comparison of PLV and RTV.

that motion video quality is subjective to some extent, and that for most applications, the 256 × 240 resolution is adequate for an application's content.

RTV and PLV files are different in quality because of the type of compression used in each process. Figure 2.5 summarizes these differences.

Still image and audio compression

Motion video is not the only media element of a multimedia application that benefits from compression, however. Still images and audio are also compressed using DVI® technology. The capacity of a 650,000,000-byte CD-ROM when using DVI® files is outlined below:

TEXT	650,000 Pages
AUDIO	5 Hours FM Stereo or 22 Hours Midrange Monaural or 44 Hours Near-AM Quality Monaural
VIDEO STILLS	5,000 Very High Resolution (768 × 480) or 10,000 High Resolution (512 × 480) or 40,000 Medium Resolution (256 × 240)
MOTION VIDEO	72 Minutes of Full Screen, Full Motion (256 × 240 resolution at 30 frames per second)
MIX and MATCH EXAMPLE	20 Minutes of Full Motion Video with 5000 High Resolution Stills with 6 Hours of Audio over Stills with 15,000 Pages of Text . . . all on one CD-ROM

The i750 video processors

In November, 1990, Intel announced the i750 PB/DB video processors. These video processors are what make the compression, and subsequent display of motion video and other elements in a multimedia application viable. The approach that Intel has taken with the i750 PB/DB processors is very similar to other components developed by Intel. The video processors have an architecture that has been designed so that developers of multimedia can create software applications today that will play on future generation processors—just as word processors developed for personal computers based on the 80286 microprocessor can be used on computers based on the Intel386 microprocessor, and so on.

The i750 video processor family has been planned so that future generations of motion video compression technology will also be enabled by these components. The first ActionMedia boards, based on the i750A video processors, are a perfect example of this. Applications built using these first-generation processors will be able to be played on new products based on the i750 PB/DB. And the i750 PB/DB are able to process other compression algorithms, such as JPEG still algorithms. This will be discussed in more detail in Chap. 17.

The system software

Operating in the personal computer environment requires that software is available to manage all of the video, audio, and other data in any particular multimedia title. DVI® applications can be developed in either Microsoft C, or by using an authoring language. An authoring language is a high-level set of software instructions that manages this data without requiring code-level program.

Typically, an authoring package calls the DVI® system software, which in turn calls the decompression software and DVI® hardware to implement the instructions. Authoring is quicker than programming, but also has limitations. A more complete treatment of the issues of programming or authoring will be discussed in Chap. 13.

DVI TECHNOLOGY PRODUCTS

Intel Corporation has created a complete line of products based on DVI® technology.

- The i750 video processor is offered to hardware manufacturers, including manufacturers of personal computers, add-in cards, consumer products such as game machines, automatic teller machines, and simulators.
- The ActionMedia Board Products are designed for developers and PC users to integrate into their PC's for development and playback of applications.
- Off-line PLV Compression Services
- ActionMedia Software Library and Production Tools

See Chap. 5, "Selecting the Delivery and Production Environment," for more details about each of these products and how they are used.

Video compression and personal computer standards

The printing press provided a type of standard for people around the world to communicate. Later, the telephone and the television did the same. Thankfully, for great artists like Keats or Mark Twain, little creative energy from a writer must be expended on thinking about how typesetting will effect a final product. Likewise, with film and video, the artists typically do not spend time thinking about how the material will be distributed internationally.

Companies involved in the creation of hardware and software for multimedia, and specifically for compressed video, are working to create compression standards, so that compressed video can be distributed internationally. The International Standards Organizations, or ISO, is comprised of several subcommittees, or expert groups, concerned with creating standards for both still and motion video imagery.

JPEG, the Joint Photographic Expert Group, is the subcommittee for still images. The standard itself bears the name of the committee. MPEG, the Motion Picture Expert Group, is the subcommittee working on motion video standards. Each of these groups will propose a compression (encoding) and decompression (decoding) format for acceptance as worldwide standards over the next several years.

There is debate about the importance of these standards. Some feel that it is important that multimedia applications use standard formats whenever possible. According to this way of thinking, standards will ensure the portability of files, and will help to build the business for everyone. Others think that hardware and software systems can accommodate multiple standards, while insulating the producer from particular file formats. In this approach, the hardware, i.e., video components, would interpret a multitude of video compression algorithms in a transparent way, so the producer would never have to be bothered with the format of any file. An analogy to this would be if the same videotape could contain clips from tapes produced in Beta or VHS, and your VCR could read either one.

DVI® hardware and software solutions provide support for the JPEG standard today. Intel's i750 video processor is microcode programmable, so that multiple types of compressed video can be played through the DVI® system. A third generation of the i750 video processors is being developed, available in the 1992 time frame, that will be able to process the amount of data needed to conform to the MPEG requirements.

Software standards for the personal computer present similar challenges. Operating system compatibility is one example of this. Currently, DVI® applications are DOS-based, since the system software for the first ActionMedia products is based on DOS. In 1991, Windows and OS/2 versions of this software will be available. Unix versions will be available in 1992. In this way, developers can choose multiple platforms, or the most appropriate platform for their multimedia application.

DVI® TECHNOLOGY SUMMARY

The goal of all of this work—compression standards and strategies, video processor manufacturing and integration, and system software compatibility—is to create a personal computer that can deliver all types of information with ease and transparency to users. We would all prefer that transferring information be as simple as dialing a phone, using a reference book index, or rewinding our VCR.

Multimedia applications on the personal computer open up access to information in business and home environments. Multimedia allows us to distribute information in a timely manner; we can localize the control of information delivery in ways never possible before. In the next decade, protocols for how this information is delivered will undoubtedly take shape, and will take on forms as powerful as the GUI that is sweeping personal computing.

The artist of this new form, the multimedia developer, can take on the role of early pioneers in other media—print and film are examples. Multimedia developers will create the protocols of this new form of information—the equivalent of the index, table of contents, the cut, fade-to-black, and search and retrieval engines—by understanding the basic capability, limitations, and features of this new information technology.

Chapter 3

Steps to Producing a DVI® Application

In Chapter 1, we covered the phases of creating a multimedia application very broadly. The design phase, production, programming, product testing, and documentation are all critical steps in the creating of a multimedia application product. The phase that is most different from typical software development is the production stage, when the actual content of the application is collected and assembled. Most of this book, and particularly Section 2, is dedicated to the details of the production phase. This chapter will give you a broad overview of the steps of production before we move into the specific details of producing motion video, audio, still images, and the software program itself.

Perhaps the best way to show you how a title is developed is to begin with a flowchart that illustrates the difference between multimedia title development and other software development (see Fig. 3.1). All software development will include a phase where the product team does an analysis of their goals and specifies the design of the product. After this, the implementation of the specification and design begins. Finally, the product is tested and documented.

The first stage of multimedia title development includes all the elements of designing an advertisement, writing a book, or producing a movie; figuring out what you want to communicate, identifying your potential audience, and creating an application treatment—probably

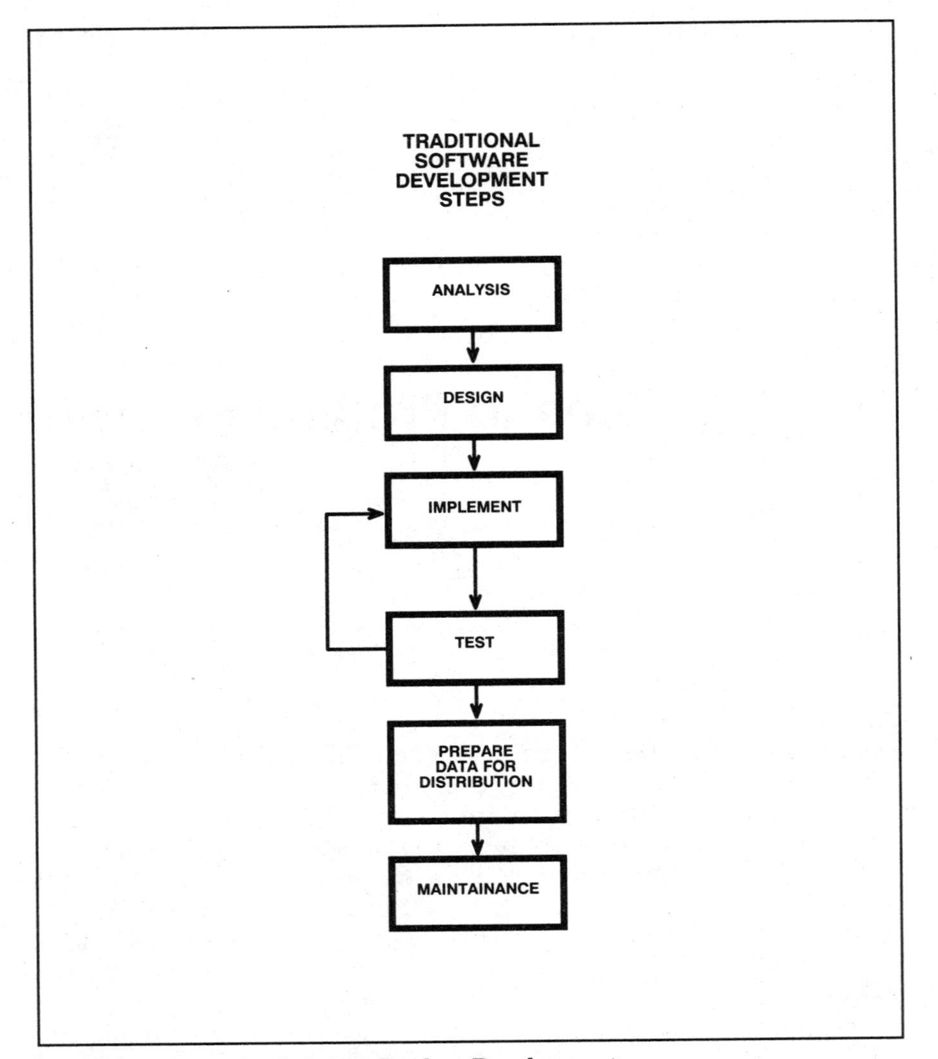

Figure 3.1 Steps for Software Product Development.

through storyboards—along with writing a technical specification. This phase is critical to the success of the product.

We will not spend a great deal of time on this phase, since it is totally dependent on the specific application you are creating, and the individual philosophies of your team in regard to design, interactivity, and the look and feel of the applications you want to bring to market. The mechanics of production are a much more standard process, and our experiences in this area can be applied across any multimedia production project.

THE PRODUCTION STAGE

The production stage is unique to multimedia application development, and may be new to anyone who has worked previously creating traditional personal computer software products. Typically, production includes a planning stage, when the logistics of the production are put into place. A script must be written. Video crew may have to be hired. A photographer may be needed. And, of course, a budget must be prepared. Chapter 6 gives you some specific suggestions and recommendations about how to best manage this part of the application development process.

After the planning stage, the actual gathering of production elements takes place. Audio is recorded, video is produced, still images are acquired, and artwork is completed. Last, all of these elements are converted and digitized into the common digital format needed to integrate them into a DVI® application.

Analog elements, such as video and audio tape, must be converted to a digital format; graphics may need to be converted from one format to another. Compression of these elements occurs at the same time they are converted to digital format. After all the elements are in the correct format, they can be used in the program as interactive "clips," or objects, that are controlled by the software. These specific production steps are outlined in Fig. 3.2.

One approach to completing a software application is to work linearly. The software and production team may finish each step in order—creating completed artwork prior to programming, for example. Another approach is to have the production steps going on at the same time that the actual software program is being built (programming or implementation phase).

In this scenario, team members work separately, but in parallel, during the production and programming stage. The production team gathers or creates required graphics, audio, and video clips while the programmer begins to write the skeleton software program. The team members come back together when the elements are ready to be integrated to complete the title.

Analog to digital

As mentioned above, regardless of whether the team is working in parallel or on a step-by-step approach, the step following production of the media elements is to digitize and compress each of the elements. This is the step that moves all of production from a variety of sources (audio

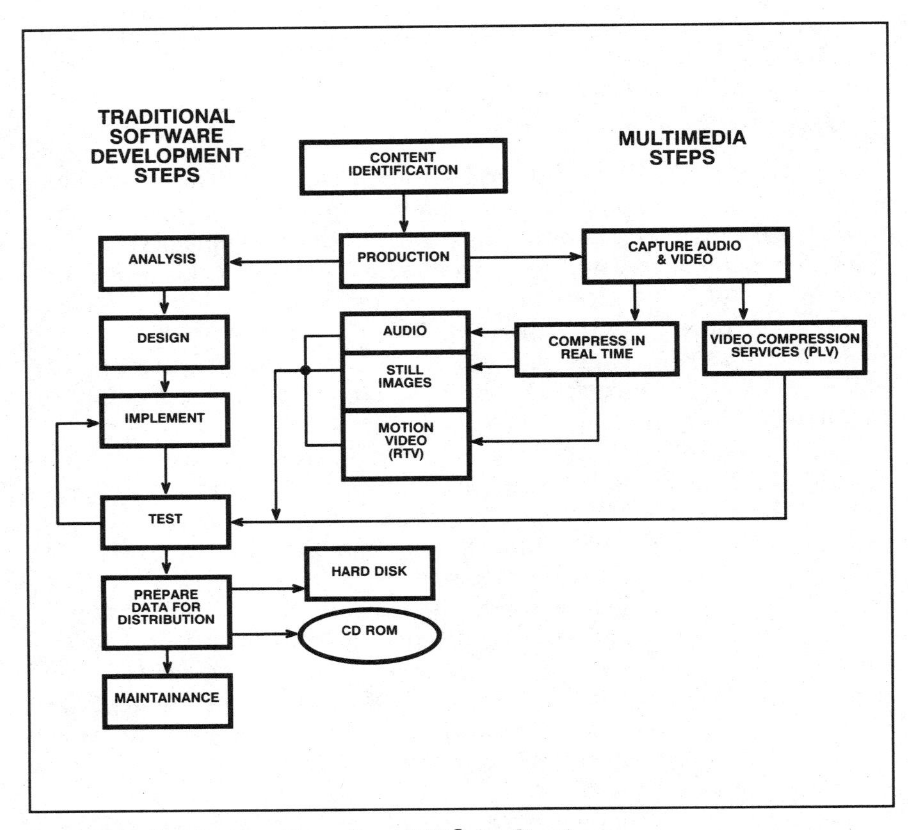

Figure 3.2 Production Steps for a DVI® Application.

and videotape, photographs, graphic artwork) into one common format. It is also the step that turns the DVI® development workstation into the central production vehicle. The DVI® workstation now becomes the center of final editing and assembly of the application.

The basic hardware configuration of a DVI® development platform includes the ActionMedia 750 Delivery Board and Capture Board, as shown in Fig. 3.3.

The capture board is used in conjunction with software to convert the analog video, still images, and audio into a digital format. The delivery board compresses the digital elements. Chapter 4 provides a complete description of this hardware and software configuration, and it is summarized here in Fig. 3.4.

As mentioned, analog-to-digital conversion in a DVI® development system is a hardware and software function. The ActionMedia 750 Capture Board is required, as are the ActionMedia Production Tools.

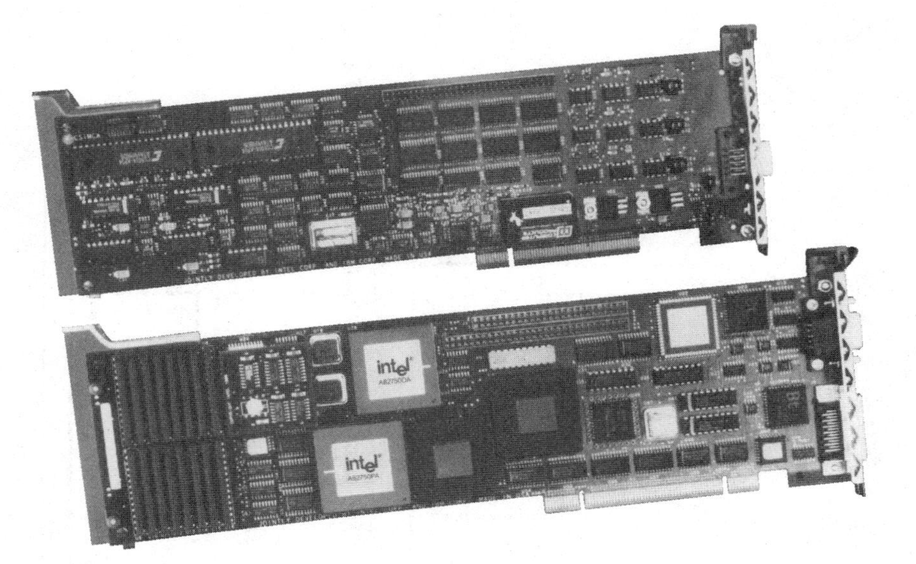

Figure 3.3 ActionMedia 750 Delivery Board and Capture Board.

Each source peripheral (camera, tape recorder, VCR) is connected to the capture board through appropriate cabling.

There are a number of software products offered by Intel and IBM for DVI® application development. For the purposes of this discussion, there are two that are critical:

- ActionMedia 750 Production Tools are the software component for capturing, compressing, and editing the audio and visual elements.
- The DVI® system software, or ActionMedia 750 Software Library, is used to create a software program in the C programming language that will manage the multimedia elements' interactivity once they are digitized and edited.

PRODUCTION SOFTWARE

The Production Tools can be classified by media type, such as audio, video, or still image, or by function, such as capture, edit, or convert. (See Fig. 3.5.)

Capture tools

Capture tools exist for still images, motion video, and for audio. Capture tools (VCapt, VRTV, and VAudRec) convert analog representations of still images, motion video, and audio into digital data and

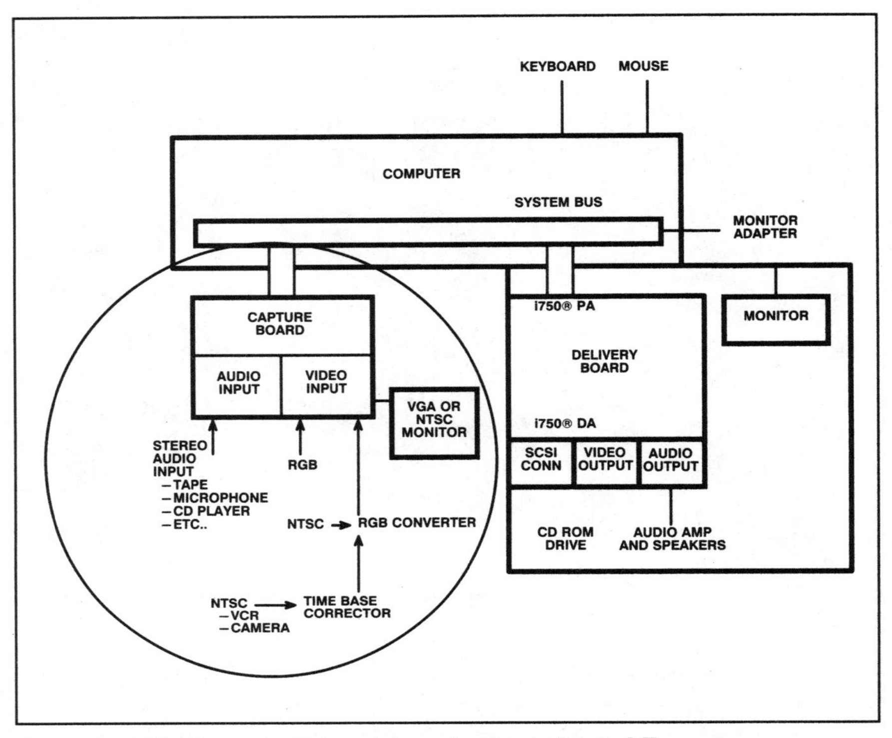

Figure 3.4 Hardware to Convert from Analog to Digital Format.

compress it. The most unique of these is VRTV, since it captures motion video and audio in real time and compresses it so that it can be stored on hard disk. Real-time compression, or symmetrical compression, is an important feature of DVI® technology, since it allows motion video to be digitized and compressed on a personal computer.

Editing DVI® files

VAvEd and VAudEd are the principal editing tools and are used primarily to adjust the length of each video or audio segment. These tools can also be used to add a new audio track to a video file, or to create a new audio-video file by editing two or more together. D/Vision, an editing tool available from TouchVision Systems Inc., is also now available to use in editing audio and RTV motion video files. This software package has a user interface that is similar to professional video editing studio equipment.

Still image files can be both created and edited, through either the ActionMedia software library or through a paint package such as Time Art's Lumena, for DVI® technology. You can use these tools to

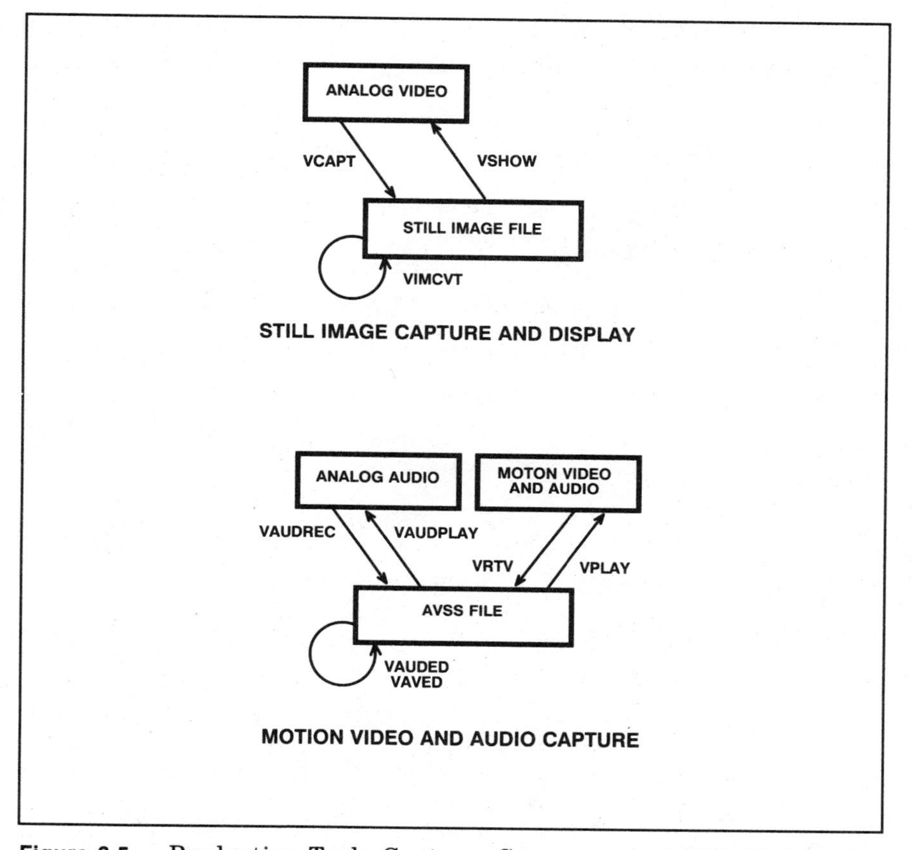

Figure 3.5 Production Tools Capture, Compress, and Edit Multimedia Elements in Two Groups: as Still-image Files and AVSS Files.

"fine tune" a still image by making it brighter or by cropping its size, zooming into detail, or by adding text or graphics to the screen.

These "edits" to the multimedia elements can also happen within the title. That is, you may want to give the end user control over the brightness of a still or the volume of an audio clip. A title may call for a user to identify certain parts of an object by "drawing" over them with graphics. These are interactions within the title, rather than edits to the media. All of these types of interactions are controlled by the software program, and are called from the ActionMedia 750 Software Library.

The conversion tool

You may want to use elements, like still images, that have been used in a different application, or that were gathered using a different hard-

FUNCTIONS OF THE ACTIONMEDIA 750 PRODUCTION TOOLS

Audio tools

VAudRec-	Records an AVSS audio file. Options: Sampling rate, frame rate for playback (frames per second), mono/stereo, volume.
VAudEd-	Edits an AVSS audio file. Options: Move in file, mark beginning and end of file, play, loop play (continuous play), copy.
VAudPlay-	Plays an AVSS file. Options: Increase or reduce filter, increase or reduce sampling rate, volume control per channel.

Still-image tools

VCapt-	Captures a still image. Options: Format (9-bit or 16-bit), resolution (up to 768 × 480), filter quality
VFisheye-	"Unwraps" a fisheye image into a 360 panorama. Options: Input and display resolutions, center clipping
VShow-	Displays a still image. Options: Image format.
VImCvt-	Converts a still image between file formats. Options: Formats (8-bit CLUT, 9-bit, 16-bit, 9-bit compressed, 16-bit compressed, 9-bit in AVSS format, LUMENA image format (PIX), 16- and 32-bit Targa image, raw 24-bit), input and display resolution, cropping.

Real-time video

VRtv-	Captures real-time video and audio to an AVSS file. Options: Frame rate, cropping

Other tools

VAvEd-	Edits AVSS files, streams, and frames; mostly motion video files.
VGrCmd-	Interactively executes graphics commands from the ActionMedia 750 Software Library.

ware or software configuration. If you use a graphic artist who works with a particular software paint package or who uses a Macintosh platform for artwork, the resulting image can be converted into DVI® format.

The conversion tool that will allow you to do that is VImCvt. In addition to converting image formats like TGA (Targa format) and

PIX, VImCvt will also allow you to convert DVI® images from one format to another. For example, if you have captured still images in 16-bit, you can convert them to 9-bit for storage and display. More details on this feature will be provided in the Chap. 10, "Capturing Still Images."

Production tool settings

One key to successful title development is careful selection of each tool's setting prior to capture, compression, and editing. These options are tightly tied to decisions about storage, memory requirements, and the quality of a particular title.

For example, a 650-megabyte CD-ROM can hold five hours of FM stereo quality audio, 22 hours of midrange monaural audio, or 44 hours of near-AM radio audio. Higher quality, whether it is higher resolution still images or increased audio bandwidth, implies a higher storage requirement.

Audio capture (VAudRec), for example, controls sound quality by offering a wide range of sampling rates, volume, and filtering. A 30-second audio file recorded at AM quality will take less storage than the same file stored at FM quality. Audio editing is used to fine-tune a file. In audio editing, you can shorten a file, fade the ending or beginning by changing volume settings, or even mix tracks to create a new sound.

In Section 2, more detail on each of these tools will be provided, with more examples of the options available, and production experiences.

THE ACTIONMEDIA 750 SOFTWARE LIBRARY

The elements of a multimedia title are its building blocks. Once the audio, motion video, and still images have been captured and edited, they form the basis of what the interactive software program manages. The software program determines how and when elements are displayed in the final application.

The DVI® programmer can create this software in two ways: by using C language routines, or by using a high-level authoring package, such as CEIT Systems Inc.'s Authology: Multimedia, Network Technology's MEDIAscript, or Cognetics Corporation's Hyperties and others now under development.

High-level authoring tools speed development through the use of a user interface, with features like pull-down menus and dialog boxes, or with a scripting language like MEDIAscript. Some access to DVI® features is lost with this type of authoring environment, so the decision to use the C language or another package must be made early in the development process.

Both of these methods access DVI® technology's system software, an MS-DOS-based software environment that makes multimedia development and delivery possible. DVI® system software is comprised of three subsystems: *Audio-Video Support System* (AVSS), graphics subsystem, and *Real-Time Executive* (RTX). These systems are needed to manage audio and video, and to turn MS-DOS into a multitasking system. Typically, multimedia applications require a software management approach that will allow more than one event to be processed at a time. DVI® system software sits "on top" of MS-DOS to manage these events in real time.

THE LAST STEPS

After the production tools are used to capture the elements, and the software program is written, the last steps of the production focus on testing and packaging the product. These steps will be very familiar to the programmer and product manager of other personal computer software packages. They are typical, mainstream activities for any software project.

If the application will be distributed on CD-ROM, there is an additional step of preparing the data, copying it to streamer tape, and sending it to a mastering and duplication facility. This step can be compared to the packaging or duplicating of software on floppy disk. You may also want to create artwork for the CD-ROM label, and if you are distributing this application commercially, you will want to have a design created for the package and documentation that will accompany the disc. The details of these final steps are covered in more detail in Section 3 of this book.

OUR EXAMPLE

The description above is a very abstract look at the development process. The best way to learn about these steps is to actually implement all of them. Short of that, we can tell you the story of one multimedia production that has actually been completed. It is a presentation that was used by an executive of a large corporation to deliver an inspiring message about taking risks and staying motivated. This multimedia presentation has also been used to teach developers like yourself about the production process.

A multimedia presentation

A multimedia presentation is quicker, less expensive, less time-consuming, and more straightforward to produce than a full-blown interactive, multimedia application. The reasons for this are fairly obvious. A presentation is usually designed to work fairly linearly, so little interactive design is needed, and is usually designed to be used

THE ACTIONMEDIA 750 SOFTWARE LIBRARY

A DVI® programming team can create application software in two ways: by using the C language, or by using a high-level authoring package. Both of these methods access DVI® technology's system software and the MS-DOS environment. DVI® system software is comprised of three subsystems: Audio/Video Support System (AVSS), Graphics Subsystem, and the Real-Time Executive (RTX).

AVSS consists of routines that control the playback of audio and video data files from RAM, hard disk, or CD-ROM. AVSS files are made up of separate streams. A typical AVSS file has two streams, audio and video, though multiple audio streams and multiple video streams may be desirable. AVSS can control the rate of play of a file, the start and stop point, the position of a file on the display screen, volume of audio, etc.

Graphics Subsystem provides graphics functions such as drawing points, circles, and lines. Special DVI® capabilities, such as bitmap manipulation, image processing, video effects, and real-time decompression of motion video, are also a part of the graphics subsystem.

RTX is a multitasking operating system that runs "on top of MS-DOS." For example, RTX is used to control the process of playing simultaneous video and audio.

Together, these three components of the system software allow the programmer to write C language calls to control how, where, and under what circumstances various elements are displayed in an application. Authoring packages for DVI® technology have been developed using the ActionMedia 750 Software Library to access the capabilities of the ActionMedia board with either a plain-language or graphical interface simplifying the application development process.

for a short, defined amount of time. Little work is needed in creating a user interface, and usually intricate programming is not needed. At the same time, many of the steps are similar to those in a full-blown application, particularly in production.

Our multimedia presentation was about 10 minutes in length. The charter was to create an inspiring, good-looking multimedia presentation to use with people who were interested in creating software for multimedia. The presentation was also to be used in a teaching environment, where the principles of production would be taught.

Content and design

The content of the presentation was driven by a few broad ideas:

- The topic should be something related to technology.
- Inspiration and success should be a major part of the message.
- Photographs, motion video, and audio should be available to choose from, since no original production was planned.

The topic chosen was flight. This came from a series of meetings where the production team discussed some ideas. Flight and flying is a technology that is based on risk-taking and success, so the first two criteria were met. There is a wealth of still images and historical footage of early flight technology. In addition, there were several short clips of modern flight that had been digitized and compressed that the team could use. Flight met all of the criteria, plus it was interesting for a number of the people on the project. (The producer and script writer both were pilots!)

After identifying the content, this project proceeded through a series of project management steps:

- Personnel and their roles were identified.

  Graphic artist—Create overall look for the presentation and some original artwork.

  Programmer—Use Authology:Multimedia to create the presentation.

  Production manager—Keep all activities on budget and on time; digitize and compress all media elements.

  Researcher—Find historical still images and motion video for use in the presentation.

  Script writer—Write the audio voice-over narration.

  Musician—Create two short original music clips.

- Budget and time line were created.
- Rough storyboards and flowcharts were created.
- Script for voice-over narration was written.

Production of the media elements

Still images and historical motion video, once identified, were digitized on the ActionMedia Application Development Platform. The historical footage, originally shot on black-and-white film at 24 fps, looked very good as RTV files. It was digitized and compressed directly from VHS tape. The stills were mostly photographs from books. They were digitized using a flatbed scanner, and then compressed using ActionMedia 750 Production Tools.

After looking at the RTV motion video files as full-screen images on a 256 × 240 screen, and as one-quarter-size images displayed on a 512 × 480 screen, all agreed that the partial screen format yielded a better-looking image. (Full-screen motion video can be copied onto a 512 × 480 screen. While the resulting image does not contain any more pixels, since it is being displayed on a higher resolution at a smaller size, the image appears crisper.) The rest of the screen display was original artwork that set off the RTV, and helped blend all of these elements from a variety of sources into one unified presentation. This same presentation style was very effective with stills, as well. (See color photographs in center section.)

All of the "elements"—RTV files, stills, original artwork—were stored on hard disk, and were named by the production manager. Names of the elements were then transferred to the flowcharts, and the programmer could begin to build the presentation following the flowchart. The PLV motion video was copied from the CD-ROM to the hard disk, was named, and was also added to the flowchart.

The audio script, which would later be recorded professionally, was edited so that all of the stills and motion video were identified by file name and coincided with the audio in the presentation. In other words, if the script called for a still photograph of the Wright brothers, it was amended to call this still by the file name assigned by the production manager. This step ensured that there was matching audio for the media elements, and was a double check that the correct elements had been identified for the application.

Missing elements were collected. Some other adjustments were made as needed. In some cases, elements were dropped. The original artwork was also adjusted at this stage. Since the audio was the first element that would have to be created in a studio, costing real dollars, it was important to finalize the script before recording.

Recording of the voice-over audio was done in a studio. Some original music was created by a professional musician. Even though audio could have been used from other sources, since this presentation would be used to teach about multimedia application development, it was decided that original production should be a part of the curriculum. The original score and voice-over were mixed on an analog sound mixing board, and then digitized and compressed on the ActionMedia Application Development Platform. This was not done as one continuous file, but in "sections" that matched and could be synched to the stills and motion video.

Programming the presentation

The programmer could now begin to put the audio, still images, motion video, and artwork together to create the final presentation. The whole presentation was to run from hard disk, so this would be the final step. Timings of the visual display and the audio were the most critical thing to complete and test during this step. Fades, wipes, and some animation were also added here, and each new special effect had to be accounted for in the timing.

File size, particularly of stills and original artwork, proved to be very critical to the application. Originally, the programmer converted all of the graphic artist's large, 16-bit images to a more compact 9-bit format available in DVI® technology. The graphic artist and production manager agreed that most of these images should stay as 16-bit after looking at the results—some detail is lost in the 9-bit format, since it is smaller in data size. (See Chap. 10 for a detailed discussion of still-image formats.) This adjustment changed timings of audio files across the board, and adjustments to the audio start and stop times had to made as a result.

The presentation was a huge success, and has been used in a teaching class, in speeches, and presentations at multimedia conferences around North America. The key to this success was a team who worked well together, who were knowledgeable about their respective areas of the development process, and who were willing to make compromises and changes as they learned critical information about the development process.

SUCCESS AS A MULTIMEDIA DEVELOPER

This story, and the stories of other multimedia applications such as those in Chap. 1, are the ones we will use to impart more information, production tips, management suggestions, and the technical knowledge you will need as you develop applications for DVI® technology. The overview here is the tip of the story, and we will dive into each area—still-image production, audio production, motion video production, and programming the interactivity of an application—in far more detail in Section 2. The information we will share is based on the stories of application development we have actually been involved in, or have discussed with other developers.

The two biggest lessons we have learned, however, are: Be organized, and be open and cooperative. There are many elements that need to be controlled and managed in any multimedia application—even with the linear, 10-minute presentation we described. It is easy to go over budget, to fall into the trap of miscommunication within the team, and to have too much intricate project knowledge undocumented. In Chap. 6, we will suggest some tools that will help you with this aspect of application production.

Second, on the cooperation front, multimedia brings together a diverse group of people. As you select people in your organization to work together, or to work with other developers, you should select those who have strong communication skills and an ability to take direction, as well as contribute direction effectively to others. This may seem like a tall order, and it is. Exceptional applications will result from exceptional people making up exceptional teams. If you want success, you must find others who share that wish, and are willing to cooperate and learn together how to achieve it.

Chapter

4

Staffing and Skills for Multimedia Production

Multimedia, by virtue of its name, incorporates multiple media. Each medium requires specific production skills and tools. In this chapter, we are going to look at the skills, and the people, you may want to have as part of your multimedia production team.

While the use of the term *multimedia* can be applied to adding audio and pictures to a business presentation, in this chapter we are going to take a broader view by assuming we are managing the development of a small production that will stand alone. This way we can examine the various skills in more detail. To show the other extreme, we will also look at the added staffing requirements for the production of a large multimedia application.

You may be one of those rare individuals with a combination of media skills that allows you to produce a complete production by yourself, or with some assistance from a small staff. Or, by necessity, you may belong to a small AV department of a corporation, or non-profit agency, with the responsibility of producing entire multimedia presentations or training materials with limited resources. With multimedia produced on a personal computer, you have the advantage that all of the tools you need for production can be installed directly on your desktop PC. But having the creative, as well as the technical, skill to use these tools is critical to the quality of the production.

The parallel can be made to desktop publishing. The emergence of great desktop publishing software didn't make everyone a good graphic designer. In fact, often it showed how graphically naive we could be as we tried to cram five different type styles into a single document! For electronic desktop publishing, the production steps and skills for creating a brochure didn't change. You still need to write copy, shoot photos, and develop graphics before sitting down with your desktop publishing software to assemble the work into a final layout. The same analogy applies to producing multimedia. You still need to shoot stills, produce soundtracks, and shoot motion video using many of the traditional techniques of audio-visual production. Like desktop publishing, at the end of the process, you can use a desktop PC to assemble all of the elements into a final production.

In this chapter, we will step through the job titles and skill sets that make up the creative and production team for a multimedia production project.

THE PRODUCTION STAFF—TASKS AND SKILLS REQUIRED

The most important resource in any multimedia production is people. People add the creativity that makes a production concept come to life, no matter how "wizzy" the technology.

The staff titles and task responsibilities for multimedia production have developed from the 100-year history of the movie business, combined with the much younger PC software industry. Because multimedia is produced on personal computers, some of the titles have changed slightly, and new roles, such as the programmer and interactive designer, have been added.

Figure 4.1 is an organizational chart showing the people required for a typical small-to-medium-size multimedia production. Keep in mind that not all of the personnel shown here are individuals. The positions represent tasks or responsibilities that can overlap. It is also useful to note that these tasks are not always full-time. Some can be done on a part-time basis, and some do not last the duration of the project. The graphic artist, for example, will complete the necessary images, and then will not be needed in the production again except to make changes or additions.

Project administration

Every project needs a management team. The following is a description of the key positions in a management team for a small-to-medium multimedia production. In general, the required skill sets for individuals on this team are a strong sense of organization, the abili-

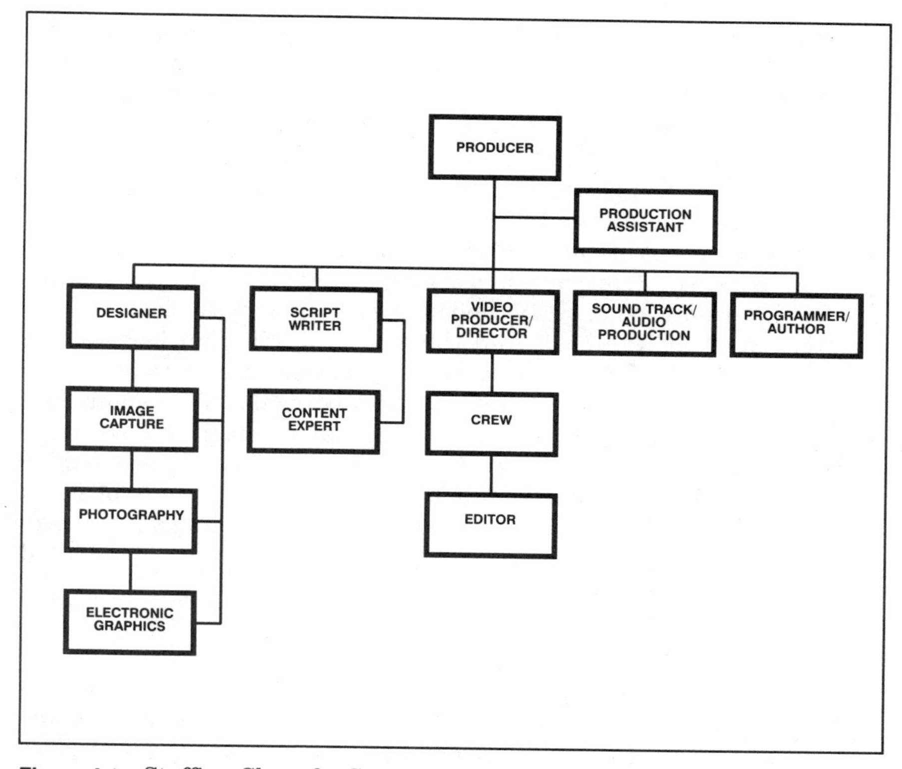

Figure 4.1 Staffing Chart for Small to Medium Production.

ty to effectively communicate with all team members, and a knowledge of media production.

Producer

The producer is the person who carries the overall day-to-day responsibility for the successful completion of the production. This includes the final responsibility for the production being on schedule, and at or below budget. The producer is responsible to make sure that the message content is effectively communicated to the target audience. The skills required to be a producer are general management skills, and a thorough understanding of all phases of the production process. One of the biggest job requirements is the ability to communicate with other team members.

The producer is the hub of a project. This responsibility includes being able to coordinate all of the people and tasks required for the production's successful completion. Interpreting written scripts and storyboard designs into finished motion video requires the ability to provide a link between the designer and writer to other team members who create the media elements. After all multimedia elements are produced, the producer is often the communications link to the programming or authoring staff, who put all of the elements together into a final multimedia application.

The producer is involved in the creative process from the very beginning. Developing the budget and managing the development of the creative concept are all part of the management skills for this job.

Where possible, experience is a critical prerequisite for the producer's job. There are no easy methods for gaining experience, as unfortunately many of the best, and the worst, production lessons are learned from the first experience. The experienced producer has a feel for how long each of the production tasks will take to be completed during the budgeting process. The producer typically knows what parts of a design will work when the artwork is created, authored, or programmed. Keeping the production process under control is an exciting and fulfilling process that comes as part of the job. The producer, while being a leader, is also supported by a team of talented and creative professionals.

Production assistant

The role of the production assistant should be emphasized. While it is commonly believed that the producer is the most important production person, much of the real work is likely to fall onto the shoulders of the production assistant. The production assistant can make or break the project. The production assistant's job is where the action is. It is where the production work is actually done.

What are the skills required? Resourcefulness and organization top the list. The production assistant is often the one who has to develop and track all of the schedules with the producer. The production assistant needs to coordinate all of the details to make sure that everyone on the project completes his or her particular task on time, and shows up on time for a video shoot or a recording session.

Depending on how the production team's duties are assigned, the production assistant may play a line role in the production, such as producing electronic artwork. Often the production assistant is in the "catch-all" position. Completing all of the details falls to this person.

For example, if a production calls for a photographer, the production assistant will be the one responsible for lining up appointments for several photographers to meet with the producer, designer, or creative director for a review of the photographer's portfolio and to assist with a final selection. The production assistant will probably then be responsible for making all of the arrangements for the photography shoot, such as booking a studio or location, selecting models, or renting props. The production assistant may also be called on to research sources of relevant file or stock photography.

The ability to maintain multiple "to-do lists" should also be mentioned, as a busy production assistant always seems to be keeping a never-ending list of things to be done for the completion of a production.

Project administrator/controller

A key task or part of the team is project administrator. Running a production is just like running a small business. During the life of the production, you will commit to contracts and purchase orders, pay vendors and subcontractors, and, of course, coordinate and collect the funding, whether it is from an outside client or from internal transfers within a corporation. Because of the number of different production elements in a multimedia project, the project administrator is responsible for monitoring production schedules, making sure that deadlines are met, and that each individual element has the proper approval of the producer, creative director, and client. Often, in a small production, the producer or the production assistant will actually handle the project administration role.

Creative team

The next major production group is the creative team. While the production team manages the project, the creative team develops all the wonderful ideas that make the production work. In general, the skill sets required are imagination, creativity, and the understanding of the medium.

Script writer

Although the title seems to define the task, in actuality the primary role of the script writer is to listen and translate. In other words, when you have a communication objective in a point-of-sale, or training program, it is the role of the script writer to listen to all of the input from people who are knowledgeable about the subject, and to incorporate this information into a final script. The script writer's listening and interpretation skills will assure that none of the pertinent information is missed. Secondarily, the script writer is an editor. A writing project starts with an overload of information. It is the job of the script writer to act as an editor to ensure that the message is delivered in a clear, concise manner.

Many multimedia productions start with the written word. A concept or treatment is often written to communicate an idea to a publisher, investor, or to management. While many of us resist writing, the benefit of a good writer on the team is the focus this person can provide. Writing forces us to organize our thoughts and think them through before we start to design the visual portions of a production.

Good multimedia script writers are hard to find because multimedia is such a new field. Script writers from television or film are used to a narrative, story-telling style. Multimedia requires a much different approach. For multimedia, the script writer should be aware of all of the multimedia elements available, and help to decide the best combination for a particular production. Often, the script writer will work closely as part of a creative team with the producer and designer. In combination, the team can work out the creative concept and production flow. The script writer develops the words, and the designer develops a storyboard of the visual ideas.

It is vital that the script writer has the ability to listen, to "sketch out" and organize ideas and concepts in words, to write in a clear and concise manner, and to provide visualization assistance for the designer. Note when we review script formats in Chap. 7, script writers will often provide notes on what they think should be visually presented on-screen.

Designer

Every production has its own unique look. The *Star Wars* movie series has a very different look from the *Godfather* series of movies; *60 Minutes* looks different from *Good Morning America.* In multimedia production, "the look" includes the background colors and texture, the typestyles, as well as the complexity of a user interface. Examples of several looks are in Figs. 4.2a, 4.2b, and 4.2c. The multimedia designer develops the general look of the production. While good design skills are the prerequisite, it helps to have an understanding of how graphics will display on screen. Working within a 3:4 ratio (the size ratio of the computer monitor), and with the knowledge of what colors work best onscreen, are all useful skills developed with experience designing multimedia production for presentation on personal computers.

Electronic graphic producer

The advancement of computer graphics software brings a lot of capabilities for electronic graphics production to the desktop. Textures, graduated backgrounds, typography, and photographs can all be combined into screen designs to complement an application. The designer may create only the graphic look or a template, and leave the work of producing the individual screens to others who can work from the initial design or template to create all the electronic art for an application.

For most multimedia programs, the creation of menus, graphics,

Figure 4.2a Example of a Graphic "Look."

and animation is the most time-intensive part of a multimedia project. Most interactive multimedia projects include a large number of menus, screens, and still images, consisting of graphics, text, artwork, and combinations of all of the above. From a program content standpoint, still images can be used to communicate a great deal of information, and should be considered a major part of your production.

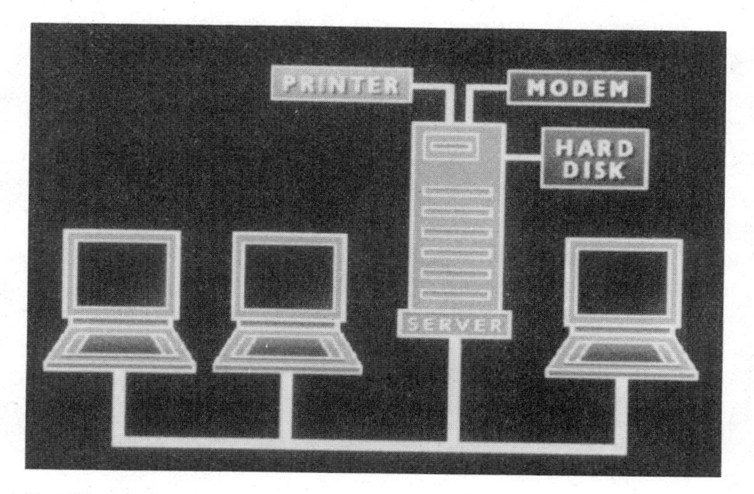

Figure 4.2b Example of a Graphic "Look."

Figure 4.2c Example of a Graphic "Look."

Image capture specialist

Many multimedia projects call for a large number of still images that need to be captured and digitized to be displayed either as the complete image or as part of the screen design mixed with graphics and text. This duty may fall to the production assistant. On large productions, it may be best to have one person dedicated to this task. The skills required are not highly technical, and can be quickly taught. It is best if the image capture specialist has an eye for cropping and scaling photos, as these decisions will typically need to be made as they capture images for a project.

Photographer

Many productions require new and original photography to communicate a point. Where existing photos do not exist, or the cost of acquiring the rights for stock photography does not fit the budget, your production team may include a photographer or someone with photographic skills who can take the necessary pictures for a production. Many producers hire a photographer on a freelance basis, contracting for several days of time as required by the needs of the project. Freelance photographers are easy to find, although the skills and specialties vary widely. Your selection of a photographer as part of your production team will depend upon the needs of the application's design.

For example, some photographers specialize in shooting outdoor exteriors that may be appropriate for a travel application. Industrial photographers are used to working in an industrial setting, and typically have equipment for a variety of lighting situations. Studio photographers work best under the controlled environment of a studio. Some photographers specialize in content areas. For example, food photographers can make food look appetizing even under the harsh lights of a studio. Choose your photographer carefully. Most have a portfolio showing their best work for your review.

Video producer

DVI® is the only multimedia format to allow you to incorporate digital, full-motion video as part of the multimedia application. Many multimedia producers come from the video production field, and have the capability of directing and coordinating a video production. If not, a project will require the needs of a video producer to plan, manage, and supervise all video shooting for the multimedia project.

Shooting for multimedia and DVI® requires a special understanding of how to shoot video for best results when digitized and displayed through DVI®. Production tips for shooting video for multimedia are covered in Chap. 11.

Sound track producer/engineer

High-quality sound adds a new dimension to the display of pictures and graphics presented through a personal computer. Many feel that the addition of audio really brings together the elements of multimedia on the PC. Adding sound and sound effects makes pictures come alive.

The special skills behind creating the audio illusion behind the images falls to a sound track producer or audio engineer working with the multimedia producer. The audio engineer's skills include knowing how to find and select music, and how to add sound effects that artfully enhance an image.

The audio engineer is also responsible for recording all narration. The audio engineer has the responsibility for converting the audio tape soundtracks into digital computer files for programming into the completed application.

Programmer or author

The programmer or multimedia author is the person responsible for linking all of the multimedia elements together into the final production. Based on flowcharts, storyboards, and scripts, the programmer or author will sequence all of the elements into the final program. The programmer works with the C programming language, calling on the ActionMedia 750 Software Library for the DVI® playback boards.

The author works with a higher level of the authoring language or tool to order and sequence the multimedia elements.

The multimedia author does not need to be a programmer, but a background in structured programming is useful in understanding branching and case statements that are often used in interactive programs. For multimedia programming and authoring, the programmer or author with a feel for or experience with multimedia has a distinct advantage. Often they will be called upon to develop innovative techniques to display images, text, and graphics and synchronize them with sound. The programmer and the author play a very key role in the final results of a production. Their role is analogous to the editor of a movie who takes all of the film and sound elements and brings them together into the final production masterpiece.

A small production team is the most efficient

In a small-to-medium-sized multimedia production, many of these roles are mixed and matched. Sometimes you will find (or you may be) the rare individual who can act as producer, designer, script writer, and with the help of an authoring language, even the programmer. Many corporations have small teams of two to three people who share the duties of a training or corporate information multimedia production. The small team can be the best solution. A small group of talented and experienced multimedia professionals can work economically and efficiently because the lines of communication are shorter.

Staff for a large application project

Some multimedia application development projects are too large for the small team. When a production grows in size, it will require another layer of management. Due to the sheer volume of production work required, additional personnel for graphics production, image capture, and programming or authoring will be required as part of the production team.

The following job areas describe the skills and tasks required for the larger multimedia production team. An alternative organization chart for a larger production is shown in Figure 4.3.

Executive producer

The executive producer is responsible for the business aspects of the production. This includes raising the initial funding for the production, as well as overseeing the management throughout the production. It is typical that the executive producer may be managing multiple projects, with producers being dedicated to each project. In the corporate environment, the executive producer could be compared

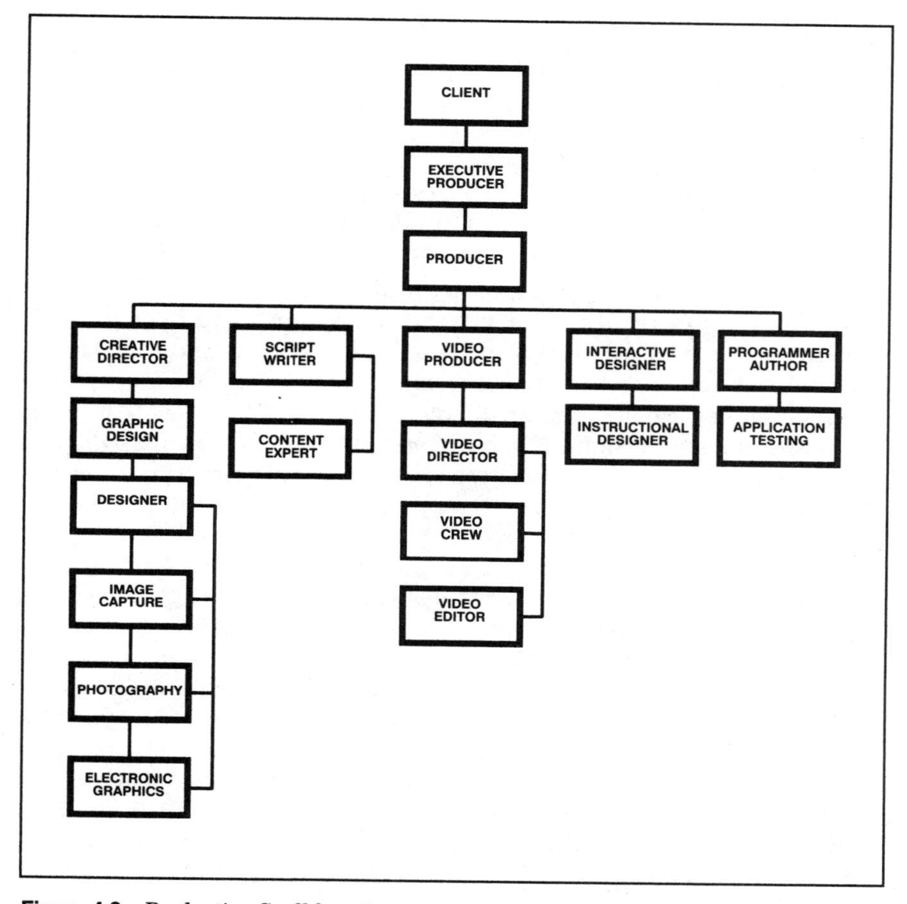

Figure 4.3 Production Staff for a Large Project.

to the department manager. The parallels are very similar, as the department manager has responsibility for budget and reporting to either the client or top management on the progress of the project.

Creative director

In a large production, responsibility for the overall look and feel of the program will often fall to the creative director. The creative director has responsibility over all creative elements to make sure they meet the communications needs of the project and are consistent with the interest of the client. The creative director focuses on the content of the production. As the executive producer is responsible for the administration of the production, the creative director oversees and manages the creative output.

Graphic designer

In a large application development project, the graphic designer will have responsibility, along with the creative director, for the overall look of the production. The graphic designer has responsibility for designing each of the onscreen presentations or a template. This may change from section to section of the application to provide some diversity within the application, but overall must have the consistency of the same look and feel. A graphic designer will design all of the individual areas of an application with this in mind.

Where do you find a good graphic designer? The best graphic designers for multimedia production are those already designing graphics for television production. They are used to working within the bounds and constraints of the 3:4 ratio of a TV screen. They also know the rules of design for screen presentation. Some colors display better onscreen than others. Onscreen text and display type needs to be handled differently for onscreen production than for the traditional graphic artist's typical print environment.

The basic skills required for a graphic designer for multimedia are similar to those required for print design. The print graphic designer is trained to build a balance into a design with space, color, and type. These same basic skills apply for an onscreen display, but from there things are different. The multimedia graphic designer needs to understand transitions and how to successively build screen with type and images. They need to know the flexibility and features of the technology so that they can take full advantage of the creative opportunities multimedia offers. Experienced print graphic designers have this basic training. For the graphic designer who is new to multimedia, you may have to help interpret the design to take full advantage of the capabilities multimedia offers.

Graphic designers are often challenged by new media. If you are reviewing a graphic designer with a strength in print design, present the opportunity of working in multimedia as a new challenge. Emphasize all the different creative and dynamic possibilities multimedia offers beyond printed media. The creative option of easily mixing graphic with photographic quality images, and the graphic tools available, give a multimedia graphic designer almost unlimited possibilities. Most designers will get a creative kick out of this, and you can develop an excellent, mutually beneficial working relationship.

Interactive designer

Designing an interactive experience within an application is a new skill that goes beyond design skills required for a typical linear videotape or movie. Designing menus, branching, and general program flow is only part of this. Good interactive design has a reasoning and meaning behind the directions the program will take. A good interactive designer has an understanding of user interface issues and navigation. Applications should be designed so that the end user feels at ease, and can move within the program as he or she wishes.

For example, in a training application, the ability to loop back to a given subject, and review information that may have been missed or answered incorrectly in a quiz, is a new technique that most producers, directors, and designers from traditional media are not used to working with. The role of the interactive designer is to look at the overall content of the script and begin to design and lay out the flow or paths of an application typically through flowcharts. This is especially important in instructional applications, where the interactive program will be designed to gauge the performance of the viewer of the program.

The interactive designer has the duty to make sure the user interface is clear, and every possible navigational path is covered. Are escape routines available, so that the person sitting in front of the system can easily exit an area? Can one see how to move forward, backward, and most importantly, how to obtain additional information on a particular subject? When the user branches to discover additional information in a remedial or hyperlink mode, can they easily return to where they were in the program?

A design must be intuitive to the user, meaning it needs to anticipate where the user may want to navigate next within the program. Every interactive designer's worst fear is that the final program is going to be completed with dead ends where the viewer is confused and cannot find the direction to proceed.

Instructional designer

Training applications often require the skills of an instructional designer. The instructional designer knows how to create an educational experience that is both meaningful to the user and achieves educational objectives set out in the beginning of the project. Interactivity allows the user to be in control. When a student is having trouble with a concept, the well-designed interactive program provides a way to get assistance and additional information on that topic. Most importantly, interactivity can motivate learners, since the experience is fun and interesting.

Instructional design is identified closely with interactive design, as these two roles on the production team are closely intertwined. The points of view, however, are different. The instructional designer's responsibility is to make sure that all of the instructional objectives of a subject are covered. When a key point is presented for the first time, there need to be check points to determine if the information has been absorbed and retained. If the user does not understand the concepts being presented, the program will have to provide additional sources of information and instruction to the end user.

You can compare this to the job of a classroom teacher or instructor. The teacher, while leading a class, is constantly watching for cues from students as to whether or not the information being taught is being understood and retained. A perplexed or inattentive look on a face, or uneasy body language, may indicate that a key instructional point is being missed. The role of the instructional designer is to project and plan for the potential of those difficult instructional points, present an alternate way of learning more about that topic, and determine when competency is reached, or when the end user should be presented with additional information.

Where do you find instructional designers? This position on the production team is a difficult one to fill, but this is changing. Many colleges now offer graduate degrees in instructional design. In addition, educational software designers, computer-based training (CBT) professionals, and good instructors from schools, colleges and universities, and corporations all have skills that can be applied to multimedia design.

Researcher

For many multimedia productions, the material required may not be readily attainable. For example, if you were assigned to develop an educational program on World War II, someone would have to work to find the pictures, film, and audio clips to build into the multimedia production.

The researcher works with the program's script and storyboards to identify sources of visual and audio material that will cover every point presented in a script. The good researcher locates and selects a variety of material on a subject to give the producer and designer different points of view to choose from.

Where do you find researchers? The key skills are a logical mind, resourcefulness, and a pleasant telephone personality. The researcher often spends more than half of his or her time on the phone convincing libraries, museums, stock picture houses, stock film and videotape houses, and other resources to please allow them to beg, borrow, or steal material for a production. You will find that many times the use of some of this material will carry a cost or royalty. You may want to consider giving the researcher the budget authority to negotiate on your behalf for usage rights or the acquisition of various multimedia elements.

Video production crew

Many of you probably have video production skills or the available resources to call when you need video production. Like the video producer-director, it is important that the video production crew understands the goals when shooting interactive multimedia. If they are not experienced in interactive multimedia, they will probably not appreciate why they have to shoot the same scene over and over again with slight variations in the dialogue for the different elements needed for interactivity.

Application testers

In our optimism during the production planning, we too often assume that an interactive multimedia production will work the first time it is programmed. This is often not the case, and is complicated by the fact that a very interactive program obviously doesn't play straight through, and each combination must be tested and tried before the application is released. On a large project, make sure you have budgeted for application testers for this testing.

Where do you find application testers? This one is interesting, as in some cases companies doing software and interactive programming use part-time high school students. Some of the most sophisticated game players (usually under the age of 14) understand interactivity better than adults, and are often used to following a flowchart to test every combination, or to do random testing of a program.

This may not be possible with all training programs. The best group to use as testers is a sample of people from your target audience. Keep in mind that you will need someone to monitor the tests and provide feedback to the production group for adjustments to the program where necessary.

Content provider or content expert

The interactive multimedia production team cannot be expected to be all-knowing. While they probably understand multimedia production for an aircraft maintenance training program, they most likely know nothing about fixing an aircraft. The content provider is the key member of the production team who supplies the information necessary to accomplish the communications goals of the production. While this is easy to see in a training application, a content provider is also necessary for almost all applications ranging from point-of-sale to history, education, and travel.

The content provider is typically not a part of the production organization. They may come from the client, or may be hired from a university or specific segment of industry. The content provider is called on heavily during the design and development of the production. During video shooting, the content provider is frequently called on to double check that the information is being portrayed accurately.

The client

One often-overlooked member of the production team is the client. Clients play a key role, as they typically are financing the production of an application. In addition, they have an inherent interest in seeing the project done. The client may also be one of the content providers for a project, or the liaison to the content providers within the client's customer group.

Where do you find good clients? Good clients are everywhere. The need for multimedia information presented on personal computers is universal. The capabilities of DVI® allow a great deal of flexibility and interactivity in presenting information well-suited to clients everywhere.

Educating a client is the best way of having (and keeping) a good client. The more they know about the capabilities of multimedia, the more inclined they will be to find additional uses for it, requiring more production. During production, an educated client can be a valuable part of the production team, since they have the responsibility to achieve the communications goals of the program. An educated client can also better understand and appreciate the challenges a producer faces in the midst of production.

SUMMARY

Managing your resources is one of the most creative tasks you will undertake while producing an application. It is like a chess game, where you start by planning your strategy. You try to anticipate how things will work out, and then drive toward the creative goals you have set. With the right attitude and frame of mind, the challenges of bringing all of the pieces together to fulfill the main objectives, and still meeting budget and hardware and software specifications, multimedia production can actually be a lot of fun.

A multimedia production requires many skills crossing several disciplines such as graphic design, photography, audio production, and programming. You need to consider each job skill and the time required to complete that task when planning a production, even if all of the work will be done by a one- or two-person staff or department.

A large part of Chap. 6 is devoted to information about managing another important production resource, the budget. Again, because multimedia encompasses so many different elements tied to your production resources, tracking and controlling the budget for a produc-

LEARNING MORE ABOUT BEING PART OF A MULTIMEDIA PRODUCTION TEAM

While multimedia production is a relatively new field, communications using traditional media is not. Colleges and trade schools around the country offer courses in script writing, graphic design, television production, and photography. The basics taught in these courses still apply for multimedia. Courses in these subjects at local schools and extension divisions to colleges and universities can provide an excellent background at a relatively low cost.

There are also a number of additional available books and resources that you may find helpful if you are interested in learning more about a particular aspect of the multimedia development process. A list of these materials are in Appendix A.

Acknowledging the need for more content providers, the manufacturers and software publishers of DVI® products offer training programs to help multimedia professionals learn how to use them and create great applications. Several multimedia Value Added Resellers and consulting companies also offer onsite training where they will send a specialist to you for a two- to three-day-or-more session to teach you and your staff how to produce multimedia applications with DVI®. A list with the names and addresses of manufacturers and Value Added Resellers specializing in multimedia is included in Appendix B. Call them for more information on the schedule, location, and cost for training programs.

tion is a critical part of the ultimate success of a project. Producing a successful application presents many opportunities during the production process for success or failure. Having control of your production staff and the budget will allow for the opportunity to create great multimedia applications. Managing your resources is a critical step to building the foundation for a successful multimedia production.

Chapter

Multimedia Hardware: Selection of the Delivery and Production Environment

This chapter describes the development environment for creating DVI® applications and the delivery system needed to play the application back. The DVI® development system is broken down into the base system and a series of subsystems for tasks such as image capture, graphics, and soundtrack production. In addition to the hardware, the software for each subsystem is described.

One of the key design considerations when developing a multimedia application is the *program delivery* or *playback environment.* We will review the options available for selecting and recommending a full-multimedia personal computer to deliver applications. The specifications for the delivery system must be established early in the planning stages, as the specifications will affect the design of an application and other decisions made during production.

THE BASE DVI® DEVELOPMENT SYSTEM

The base development system starts with the selection of a personal computer for installation of the DVI® ActionMedia delivery and capture boards. The ActionMedia boards were designed to take full

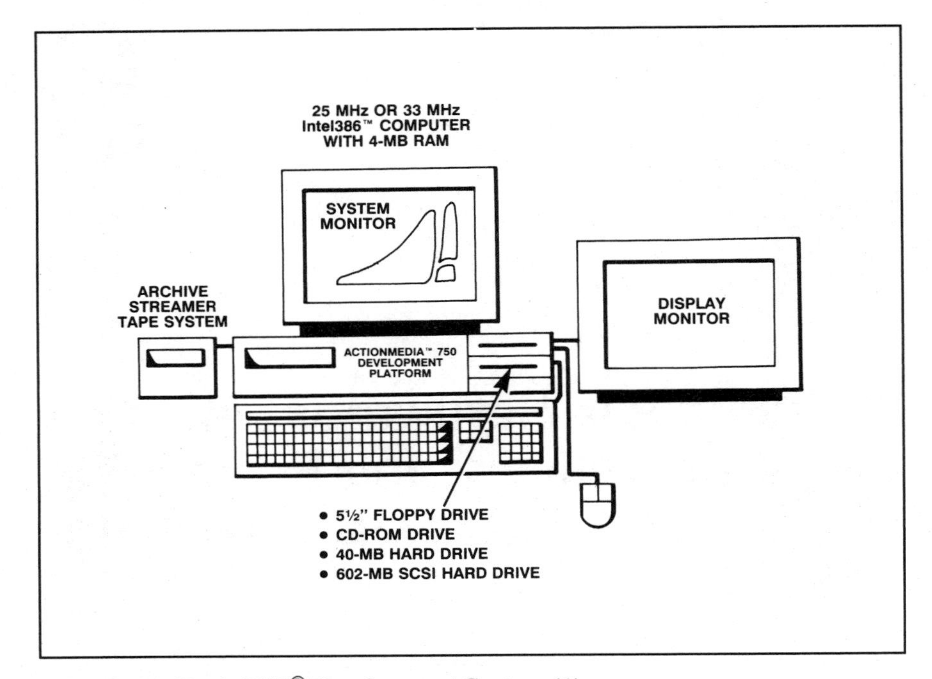

Figure 5.1a Basic DVI® Development System (1).

advantage of computers equipped with an Intel386 and Intel386 SX microprocessor. This extends from computers based on the economical 386 SX microprocessor to computers equipped with Intel's 25 MHz and 33 MHz microprocessors. For production and application development, either of the 25 MHz and 33 MHz Intel386 DX microprocessors offers the optimal performance. The high-performance processors save time during the development process when working with many of the development tools, or when compiling C program code. Figures 5.1a and 5.1b illustrate the components in a base development system.

For application development, a number of computer systems have been tested for best results and compatibility with all of the DVI® application development tools and peripherals. The reseller of your ActionMedia multimedia products can advise of you of those computers that have been tested and approved for compatibility.

Intel markets a development platform based on its Intel386 microprocessor, appropriately called the ActionMedia *Application Development Platform,* or ADP for short. The ActionMedia ADP comes equipped with the ActionMedia delivery and capture boards already installed. A 40-megabyte hard drive is also installed and con-

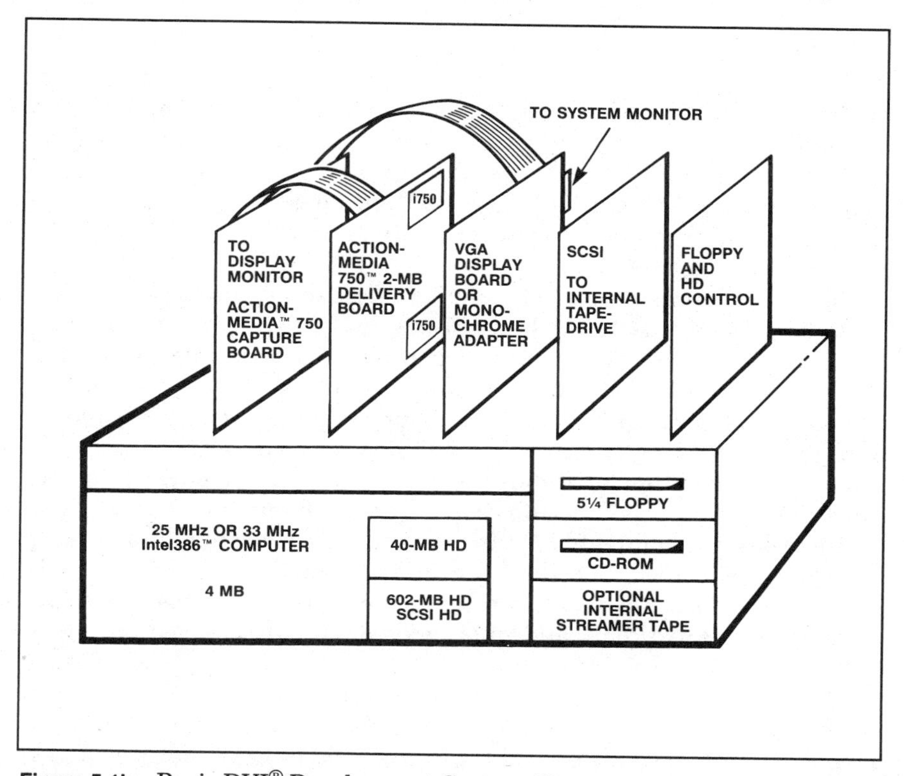

Figure 5.1b Basic DVI® Development System (2).

figured as the C: drive. The 40-megabyte drive is used for the programming software tool and storing the compiled executable code for your applications.

For development on PCs based on the Micro Channel architecture, the IBM PS/2 Model 80-32, equipped with ActionMedia MC boards (for Micro Channel architecture), is the recommended environment. The PS/2 Model 80 comes with a 320-megabyte hard drive and four megabytes of memory.

RAM memory

It is recommended that the development platform PC be equipped with four megabytes of RAM memory for DOS and EMS (*Extended Memory Standard*) support. If the development platform will be used extensively for creating graphics with Lumena, eight megabytes of system RAM should be considered.

Disk storage

A very large, fast hard-disk storage system is a convenience for multimedia production. Still-image, audio, and video files for multimedia are typically very large. Most interactive multimedia applications contain a large number of these files. In addition to the files for the final application, working files created for icons, templates or variations of edits, or elements of work in progress will be stored on the hard disk during production.

Many developers intend on having their applications distributed on CD-ROM. A CD-ROM has the capacity for storing over 650 megabytes of data. If you intend on producing applications for distribution on CD-ROM, you will want a hard disk storage system at least as large as your planned distribution medium so that you can prototype and test your complete application. The alternative is to work with a smaller hard drive and break a project down into sections. The individual sections can then be transferred to backup tape as they are completed. This may reduce the costs of purchasing a large hard drive, but it is not convenient. If you choose this method, you will not be able to test your entire application in its entirety from hard disk, and must wait until it is pressed on a CD-ROM.

Having a hard disk approximately the same size as a CD-ROM will allow you to prototype an application in its entirety directly from the hard disk. Most developers feel that they need a larger hard-disk system or multiple hard disks to store their application and all of their working files for one project or multiple projects.

Hard-disk storage systems are available in a variety of sizes. Most DVI® developers have selected the Seagate 602-megabyte SCSI (*Small Computer System Interface*) drive as the primary multimedia data storage hard disk. Some are also working with the Seagate 1.2-gigabyte SCSI drive. For the PS/2, developers can install one or two IBM 320-megabyte drives to simulate the storage space of a CD-ROM.

The large hard-drive system typically uses an SCSI interface and requires its own SCSI controller card and software. SCSI offers the convenience of interfacing up to seven SCSI devices from one adapter card. In the typical DVI® development system, this capability is used to interface to a streamer tape system for receiving digitized video from compression or for sending your data to a CD-ROM pressing facility.

Streamer tape system

The Archive Streamer Tape system is used as a low-cost medium for transporting and off-line storage of the multimedia data files, or parts of a complete application. It is used by the DVI® Compression Services Facility as the medium to return digitized files for your compressed PLV video. The Archive's DC600A tape cartridges are also accepted by most of the CD-ROM pressing plants for mastering your application to CD-ROM. The tape cartridges have a capacity for storing 150 megabytes of data. One application may require several tapes. The streamer tape system is also useful for periodically backing up your working data from your hard-drive system.

Even though it would appear that over 600 megabytes of hard-disk storage is plenty, you will be surprised how quickly a production will fill it up. The steamer tape system can be very useful to archive working files onto a tape cartridge. Saving your work in progress throughout the production is very helpful, as many times you will need to go back to one element of your production to correct a screen or reedit an audio file.

Establishing procedures for backing up your data is smart. The risks of losing several hundred megabytes of multimedia files could be devastating to a production's progress. We have found that a backup library organized with a file name index will allow you to easily find the material you need in case of a hard-disk failure.

Mouse

The mouse has become an industry standard for a pointing/input device, especially with the graphical interfaces of many applications programs. Many DVI® applications incorporate a mouse as the primary input device. In addition, the authoring and graphics creation programs for DVI® application development all utilize the mouse as their primary input device.

Display monitors

During application development, your system must be able to act as both the development system and a simulation of the playback system, so you can visualize the end results of creating screens and programming. It is also helpful to have a program monitor displaying system commands on-screen as they happen. The DVI® development environment supports two monitors for this purpose. One monitor is connected to either a monochrome or VGA display adapter as the primary computer system monitor. The other monitor plugs into the DVI® video output. This allows viewing output of the ActionMedia board only, or the combination of the VGA display and the DVI® display can be overlaid or keyed to appear on the same monitor.

The final monitors' setup for your development system offers several options which are dependent on the configuration of the final delivery system for your application. The DVI® ActionMedia display system has multiple modes of display. An application may be designed to play back through standard RGB NTSC monitors or a video projector at a scan rate or 15.75 KHz. The ActionMedia display boards also have the capability to display at a resolution and scan rate compatible with VGA monitors, 31.5 KHz. In VGA mode you have the additional choice of overlaying, or keying the output of the VGA card and mixing it with the signal of the ActionMedia display board for presentation on one monitor. This is the recommended configuration for most delivery systems.

To mix the VGA graphics display and the DVI® signal, a compatible VGA adapter must be installed in the PC. The ActionMedia 750 Display Board Installation Guide (documentation included with your board) includes a list of compatible VGA boards.

The VGA display signal is sent to the ActionMedia 750 display board through a cable between the feature connector on the VGA adapter and the ActionMedia 750 display board. Under software control, the VGA signal can be keyed over the DVI® signal. One example of how this may be used is when the time or score for an interactive game is calculated by the computer and continuously displayed on a single monitor. There are other examples of applications that mix the video output from the computer's VGA adapter with the ActionMedia 750 display in Chap. 9.

This feature gives you several options during application development. Some developers will design their applications for display only with the DVI® information. During application development, they typically prefer to use the computer's display to monitor their code. They have the choice of using either a monochrome or a VGA display for the system monitor.

If your application design calls for the use of the two monitors during delivery, or a single monitor for display of the VGA and DVI® information on the same screen, the VGA monitor is recommended.

For application development with Ceit Systems Authology: MultiMedia authoring software or Time Art's Lumena for DVI® computer graphics paint software, a two-monitor configuration is required. Both programs support a monochrome or a color VGA system monitor display. A color monitor is preferred, as both programs make good use of color to highlight different windows and commands.

CD-ROM player or drive

The CD-ROM, with its capability of efficiently storing up to 650 megabytes of data, is ideal for multimedia. DVI® motion video

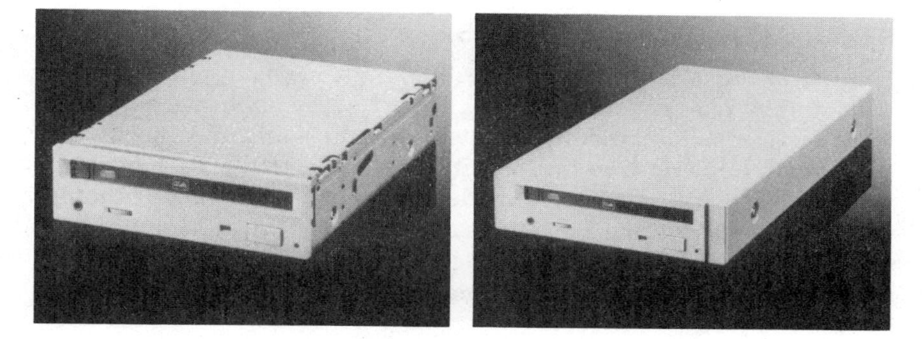

Figure 5.2 Internal and External CD-ROM Drives.

requires about nine megabytes of data storage for every minute of full-motion video. Compressed 16-bit image files require 136K of storage for each image. The CD-ROM can be pressed and produced for a manufacturing cost of less than $2 per unit. A CD-ROM drive can be purchased for between $500–700.

For the development system, the application developer has two choices: either an internal CD-ROM drive, or an external drive system. The internal CD-ROM is the form factor of a half-height, 5¼" storage device similar to the size of a floppy-drive system. When installed in a typical development system, it can be conveniently located just below the floppy drive.

The ActionMedia display board includes an SCSI interface to connect to the CD-ROM drive. For an internal CD-ROM drive, an SCSI cable is run directly between the internal SCSI interface plug on the top of an ActionMedia display board and the back of the internal CD-ROM drive. In Fig. 5.2, you can see the differences between internal and external CD-ROM drives.

The external CD-ROM drive system configuration is available for those development systems that either do not have the room for the half-height internal CD-ROM drive, or prefer the option of sharing a CD-ROM drive between several computers. For many productions, the CD-ROM is not required during the production process, as most of the data will be stored on the computer's large SCSI hard drive. The external CD-ROM drive system allows the drive unit to be conveniently moved from computer to computer and connected when needed.

The external CD-ROM drive's physical size is larger, due to the need to also include a power supply in its case. The internal CD-ROM drive takes its power from a cable to the power supply of the application development computer.

OTHER DATA STORAGE OPTIONS

Because of the large number of files in a multimedia application, your data storage plan is a very important part of the application development environment. Deciding what information and files need to be constantly kept on-line and available versus those that could be stored off-line will optimize your development environment. Another factor for consideration is the need to transport files among several specialized workstations during the application development process; for example, the artist's workstation to the programmer's or authoring workstation.

Bernoulli, WORM, and erasable

Removable media is the ideal solution to this. Iomega makes a system called the *Bernoulli hard drive.* The Bernoulli removable cartridges hold 90 megabytes of data. This size works well for use with artist workstations or for holding portions of the working files for an application, particularly still images and application code. The Bernoulli system has two configurations with either a single-drive unit or a double-drive unit which will hold two 90-megabyte cartridges on-line. The cartridges are accessed just like a DOS hard-drive device, and have their own interface card for use in the development environment. Small motion-video files can also be stored and played back from Bernoulli cartridges.

Optical storage using laser technology has evolved to allow systems which allow you to record between 200 and 500 megabytes of data on a removable cartridge system. The technology employs a laser like a CD-ROM to read the data arranged as a series of reflective pits on the disk surface corresponding to the computer's binary off- or on-base language. The laser's power is increased to record data.

Special media with thin foil in between the plastic or glass surface is used. The higher power of the laser used in recording can burn pits into the reflective foil. This burning process cannot be reversed, resulting in *Write Once Read Many,* or WORM, technology. WORM drives continue to drop in price, and are available today for between $1500 to $3000.

More recent improvements in optical media include recording media and technology that permit you to record, erase, and rerecord data on a disk cartridge. The erasable, high-capacity, magneto-optical disks are available from a number of manufacturers. Because of the different interface systems, each must be individually tested with your DVI® application development system for compatibility. The magneto-optical systems store between 200 megabytes, and in some cases, over a gigabyte of data. They are a significant investment for the development system, running over $5000. This is offset by the convenience of being able to take cartridges and individual projects off-line and on the system as needed for a production.

There are a number of other high-capacity data storage options appearing on the market. Many of these are ideal for multimedia application development. If you select another device, you will want to make sure that it has been tested with your DVI® development platform for compatibility and real-time playback. A removable cartridge system can always be used for off-line archive storage. In this case, data can be copied directly onto your hard drive for application development and production work.

LOCAL AREA NETWORKS

The ideal environment for a large production group is to have all development workstations connected through a *Local Area Network* or LAN. With a LAN, all of the planning and management files for a project, as well as the individual elements such as still images, menus, buttons, and audio and video files, are available to all members of the development team from one central location or server. This gives all team members access to the most current version of a particular file.

Several LANs have been tested with DVI®. Some will play back motion-video files and complete applications without the need for additional software. For others, software is available for LAN playback of motion-video files without pausing or hesitation.

VideoComm

ProtoComm, a company in Trevose, Pennsylvania that specializes in communications software and DVI® software development, has developed VideoComm, a driver software for a LAN server to coordinate the special needs of DVI® multimedia applications. If your goal is to play the application back directly from a server, this software will be useful. If you only want to copy files as needed onto your application development system's large hard drive for playback, you will not need this special software.

Memory requirements for LAN

Installing a LAN as part of your application development environment has many advantages, yet offers another challenge. LAN software introduces another layer of software over DOS. DOS is designed to reserve 640K as the main memory space. Sharing this memory are the application programs, DVI® software drivers, SCSI drivers, and other software drivers that need to be part of the system. The result is a tremendous reduction in the available memory space to run applications.

One solution to memory constraints is to use a software utility called "386 to the Max" from Qualitas. "386 to the Max" allows some of the drivers to be loaded into high memory above the 640K base memory space of your application development platform. This high memory is an ideal location for VGA drivers, LAN drivers, etc. "386 to the Max" software redirects the software code to know where to look for these drivers. To install all of the different peripherals you may want as part of your development platform, you will have to experiment with the specific loading order and location of the drivers so that everything works within the appropriate RAM space of the DVI® application development environment.

POTENTIAL HARDWARE CONFLICTS

In addition to the careful installation of your software, you will also need to consider how all of your hardware is installed. The architecture of a personal computer is designed to work with a series of software interrupts for control of various peripheral devices. There are a total of 16 available interrupts. Some are dedicated to existing peripherals, such as the hard-disk drive.

As you continue to add various peripheral items, such as an interface to a scanner, local area network adapter board, second or third serial port, etc., there is the risk of running out of available interrupts on your system. The layout of interrupts also has to be carefully planned when you add a new peripheral to your system. For example, many peripherals are shipped already configured for installation to Interrupt 3. In most systems, this is typically available. This may work fine for the first device, but when you go to install your next peripheral, preconfigured for Interrupt 3, a conflict will result, and the system will not work. Your reseller, VAR, or the dealer who sold you the components for your application development environment should be able to help you with the proper installation of your system. Software utilities are available to find what interrupts are available, or to determine if you have a potential conflict.

IMAGE CAPTURE SUBSYSTEM

Still images carry a very big role in multimedia for personal computers. The old adage that one picture communicates a thousand words sums it up.

There are a number of tools and peripheral products available to

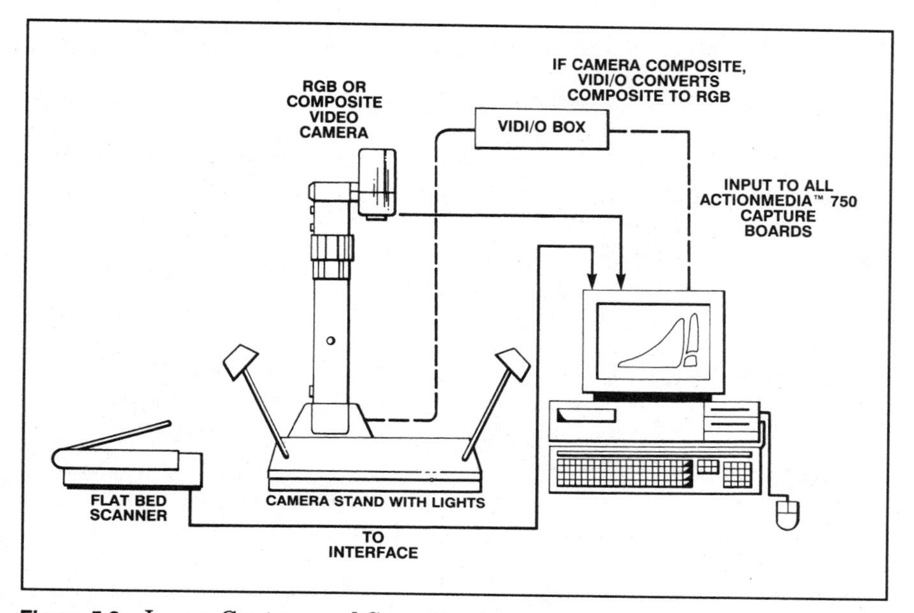

Figure 5.3 Image Capture and Scanning Subsystem.

bring this capability to the ActionMedia application development system. The Image Capture Subsystem for your DVI® development system is illustrated in Fig. 5.3. Your choice of equipment and software will be based upon your preferences, budget, and source material used for image capture.

Video camera

A video camera, mounted on a camera stand with lighting, is one of the most versatile means for image input. A camera stand is used to stabilize the camera and the image, and to provide the best possible lighting conditions. The video camera is connected via an RGB cable to the RGB inputs on the ActionMedia 750 Capture Board. The best results can be obtained by using a video camera that outputs separate and discrete RGB, or Red, Green, and Blue signals.

A variety of RGB cameras and lens combinations are available at prices varying from $1500 up to broadcast quality cameras in the $20,000–40,000 range. In most cases, a camera and lens combination including camera stand or tripod for less than $10,000 will suit your camera stand or tripod image-capture needs.

For camera stand video capture, the lens must include a macro mode to allow the lens and camera to be close to the subject material on a camera stand. Many developers like to work with a video camera and lens on a camera stand for the flexibility it offers for cropping and previewing the image to be captured.

While many applications will require image capture of a large quantity of photographic prints, artwork, or three-dimensional objects on a camera stand, the video camera can also be used on a tripod for product shots or tabletop image capture of products and three-dimensional objects. Treat the subject matter just as though you were photographing it with a film camera, paying particular attention to the design of the lighting to fit the style and artistic direction of the application.

Video cameras offering composite or S-video output can also be used when the signal is converted through either a Truevision VIDI/O converter or through some time base corrector. A VIDI/O converter does real-time conversion between composite, S-Video, and RGB signals. S-Video is a system that separates the color signal into two discrete color components for better color reproduction.

Composite video requires a simple two-conductor cable. The trade-off is lower quality and the potential for video noise when using a composite video output from your camera. For low-budget image capture, a "camcorder" type system can be used that creates a composite output. This output is fed into a VIDI/O converter box (for converting the signal to RGB) and directly into the ActionMedia capture board. When assembling your camera system for image capture, you may want to run tests of the combinations of options available to satisfy your application development quality needs and budget with the proper equipment.

Flatbed scanner

The camera offers a convenience for previewing, cropping, and capturing images. However, for the highest quality image capture and digitizing of flat objects such as photographic prints, a flatbed scanner is best. A flatbed scanner works by moving a bar of image sensors and a light source under a glass platen where a print is positioned, capturing the image without the distortion of a lens or the variances of lighting. Many scanners were designed for desktop publishing and offer a resolution of up to 300 dots per inch (dpi).

While this is excellent quality for an image destined to be printed, it is overkill for multimedia presentations that will be presented on a video screen. Most video monitors have a maximum resolution of 50–75 dpi. Scanning images at a higher resolution creates larger files, takes longer, and results in data that will be discarded when the image data is converted to a DVI® image file format.

Most flatbed scanners only capture the image into a series of colored dots in a digital stream of data that must be formatted by interface software. Time Arts' Lumena for DVI® includes scanner interface software for flatbed scanners, available from Howtek and Sharp. Both companies offer other scanner software for capturing scanned artwork and photos into a file format such as the Targa file format (TGA) that can easily be converted by using a utility called VImCvt, included with the ActionMedia 750 Production Tools software.

Sony offers a new type of scanner that outputs an RGB, S-video, or composite video signal. This scanner can be used to input directly into an ActionMedia capture board. It requires no interface software, and the connections are as simple as plugging a video camera into the capture board. The advantages are the enhanced quality of the scanner video sensors passing directly over the flat art or photograph without the distortion of a video lens.

Slide scanner

Several slide scanners are available. Lumena for DVI® supports the Nikon slide scanner directly from the Lumena software interface. Using Lumena, the image can be scanned directly into the paint software for touch-ups, the addition of type, or is used when building a series of images.

The Nikon scanner offers resolution of up to 4000 × 4000 dpi with very high precision and accordingly carries a very high initial price. It was designed primarily for desktop publishing. Like the flatbed scanners, the maximum resolution for screen presentations can be set to 75 dpi, resulting in a smaller image data file size for storage and faster scan time.

Other interface software is available from both Nikon and third-party software developers to support the Nikon scanner. If you choose to use this software, slides can be scanned as TGA files, and converted into a DVI® file format using VImCvt in the ActionMedia 750 Production Tools software.

COMPUTER GRAPHICS SUBSYSTEM

The ActionMedia display board offers a video processor and video memory storage space for still images as well as motion video. Using Time Arts' Lumena for DVI®, text pages, menus, buttons, photorealistic images, and background textures can be manipulated to create the various screens for an application. Lumena for DVI® technology software is an integral part of the graphics subsystem.

When using a professional tool like Lumena, an accurate drawing

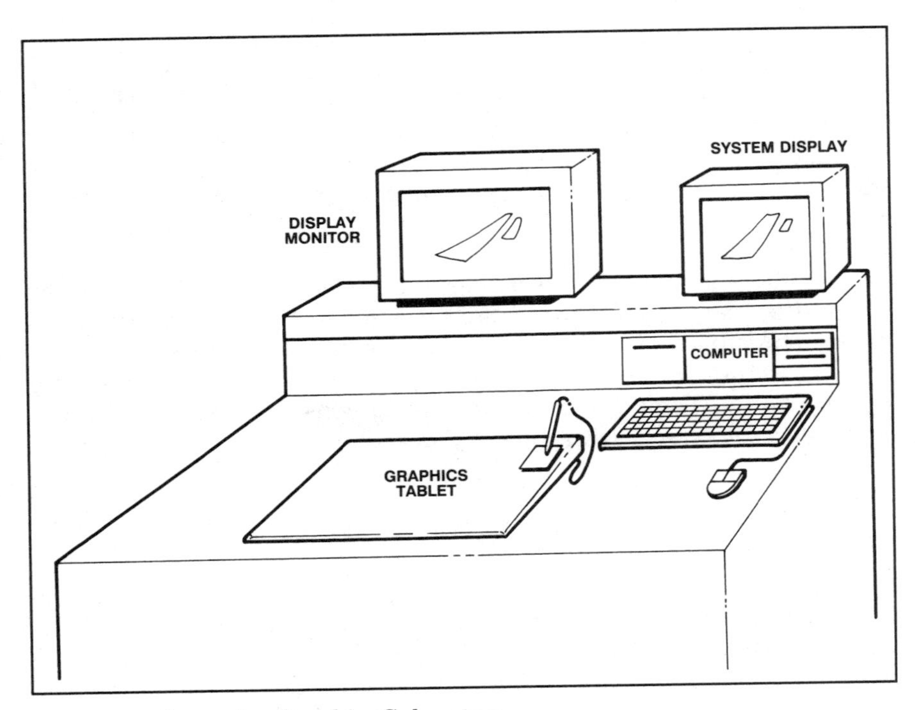

Figure 5.4 Computer Graphics Subsystem.

device is recommended. The mouse does not do the job for detailed graphics and fine still-image manipulation. In place of a mouse, most application developers use digitizing tablets (Fig. 5.4) from either Wacom or Summagraphics. Drivers are available in the Lumena software to interface directly to the digitizing tablet. Creating graphics is covered in more detail in Chap. 9.

AUDIO PRODUCTION SUBSYSTEM

Sound is not always immediately recognized as a major contributor to a multimedia presentation. Many times we think the pictures are central, when actually the soundtrack makes the pictures come alive. As a result of the popularity of the MIDI interface for electronic instruments and home recording in the music industry, a number of audio production tools are available at economical prices that allow you to create your own soundtracks for audio capture into a DVI® application development system.

Adding your own soundtrack to multimedia production can be as simple as connecting a CD player directly to the audio inputs of the

TARGA PC WORKSTATIONS FOR CREATING MULTIMEDIA APPLICATION GRAPHICS

Many multimedia application developers have access to computer graphics systems equipped with Truevision's Targa display boards and graphics creation software. The Targa-based system's output image file format is 512 × 480 pixels. Targa image files can be easily converted from their native Targa, TGA file format to DVI® file formats using the VImCvt software utility included in Intel's ActionMedia 750 Production Tools software. One of the most popular graphics creation software programs, Lumena from Time Arts, has been ported to run on the ActionMedia delivery board.

There is a wide selection of other graphics-creation software and utilities available to support artists working with PC systems equipped with Targa display boards. In addition, there is an extensive number of computer graphic artists who have workstations equipped with Targa boards and are skilled with either using Lumena or other graphics creations software such as Truevision's TIPS, AT&T Graphics Software Lab's RIO, Topaz, or Mathematica's Tempra. Images created on a Targa system with any of the above packages can be easily converted for presentation in a DVI® multimedia application using the VImCvt file conversion utility included in the ActionMedia 750 Production Tools. By using VImCvt, you have the flexibility of working with an artist who can deliver electronic graphics artwork in the TGA file format for use in your DVI® application.

USING THE MACINTOSH TO CREATE DVI® GRAPHICS

Many graphic artist professionals prefer to work with the Macintosh platform for creation of computer graphics. Images created on a Macintosh computer can be converted for display on an ActionMedia system through the following steps.

After the image is created and saved, it can be converted to the Targa file format on the Macintosh by using the Adobe Photoshop image-processing software. Photoshop will accept a variety of software image file formats popular in Macintosh environment including TIFF, PICT, PICT2, and others. Photoshop also supports a number of the PC computer graphics formats for input and output.

Once an image file is loaded into Adobe Photoshop, there are a number of image-manipulation tools available for enhancement and corrections. When the image is completed, you select the output file format. Select the TGA file format for the PC. Using either the Macintosh Superdrive or Bernoulli's File Transfer program, convert the file from a Macintosh data format to the PC's native file format. Next, load the image file into the PC and convert it from the Targa format to the appropriate DVI® image format using VImCvt.

This is a very viable option when images were created or already exist on a Macintosh system. It is also a useful technique when working with an artist whose style is appropriate for a production, and is needed for application development, but uses a different system.

ActionMedia capture board. With the ActionMedia capture board and the VAudRec software included in the ActionMedia 750 Production Tools, clips of music can be captured, digitized, and the digital audio file can be added to your application. (See Chap. 12 for more detail on audio production.) A complete soundtrack production system is illustrated in Fig. 5.5.

Adding music to your stills and menus is a nice enhancement. Narration adds depth and more specific information to support images of the application. Low-cost multichannel audio recorder systems are available to record from four separate channels directly to one or two channels on a video cassette system. These low-cost audio mixers and recorders are available for less than $500 and up to $2000. Professional high-quality results can be achieved by using the multiple channel capabilities to layer or build a sound track. Using four available audio tracks as an example, background music can be mixed with narration and sound effects to create the effect, mood, and information appropriate for your application.

A microphone can be plugged directly into the mixing panel for recording narration or new sound effects. The output from synthesizers or other electronic instruments can be plugged directly into the mixing panel for adding sound effects or synthesized background music to a soundtrack.

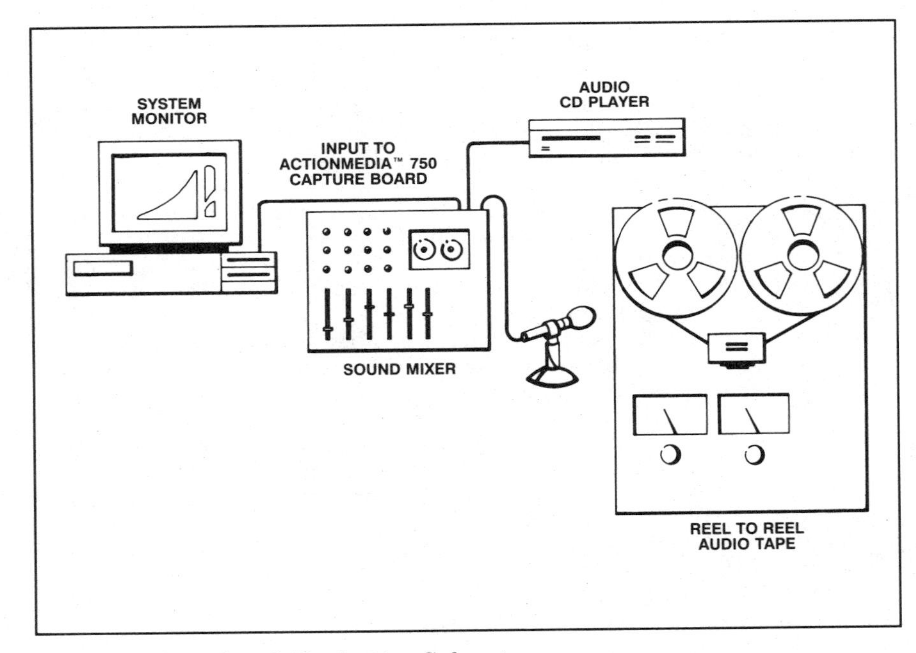

Figure 5.5 Soundtrack Production Subsystem.

MOTION-VIDEO CAPTURE

DVI® offers the unique capability of capturing motion video in real time for insertion in an application. This is called RTV, short for *Real-Time Video.* Video for an application can be captured and digitized with the ActionMedia capture board either directly from a camera or from a professional ¾″, or VHS video cassette recorder. The quality of the RTV image capture will vary, based on the quality of the video input.

Because of the differences in timing and synchronization between video cassette recorders and the ActionMedia capture and display board, a time-base corrector is used to improve the quality of the video signal. In addition, the time-base corrector offers an RGB output signal for direct connection to the ActionMedia capture board. Time-base correctors can cost between $1500–$10,000. A time-base corrector is highly recommended for capturing video or stills from video tape. The video capture subsystem is illustrated in Fig. 5.6.

SOFTWARE FOR DVI® APPLICATION DEVELOPMENT ENVIRONMENT

Intel offers several software products for DVI® application development. The ActionMedia 750 Production Tools software is a series of programs that allows you to capture and compress motion video, still images, and audio. For example, a software tool for the capture of RTV is included for real-time capture of motion video directly on your DVI® application development platform. Software tools are also included to assist with CD-ROM layout and preparation for mastering. These tools were introduced in Chap. 2.

For applications which will be developed in the C programming language, the ActionMedia 750 Software Library is available. The C-level library allows developers to create programs that call on the unique capabilities of ActionMedia board products. The library includes microcode routines that are loaded on command directly to the ActionMedia boards for video display and audio display. For example, the library includes C-level routines that display still images at various screen resolutions and sizes. Additional routines are available to manipulate the video bit map, for example, to create a wipe effect. The C library provides the ultimate flexibility with DVI® board products. All of the major routines can be called and combined together into a completed application. The ActionMedia 750 Software Library and Production Tools are discussed further in Chap. 13.

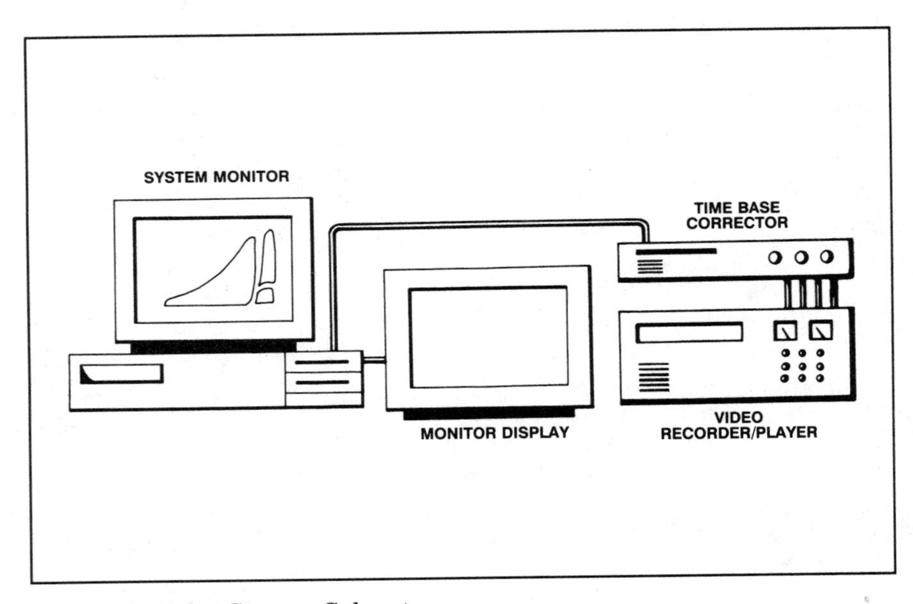

Figure 5.6 Video Capture Subsystem.

APPLICATION DEVELOPMENT AND PRODUCTION TOOLS

Computer graphics software

To aid in the development of graphics and screens, Time Arts has enhanced its popular Lumena computer graphics creation and paint software to work in the DVI® application development environment. The Lumena program was originally developed to work on PCs equipped with Truevision's Targa and Vista graphics display board products. The design and commands available on the Targa versions of Lumena have been carried over to the DVI® application environment. This is a significant advantage, as there is a sizeable group of artists and designers worldwide that learned to use Lumena in design schools, colleges, universities, and graphic design firms.

Lumena for DVI® technology offers several advantages over the Targa version. Because of the graphic processing power of the i750 video processor, Lumena for DVI® technology utilizes microcode to improve the performance of a number of significant artist tools, such as the airbrush effect. Many artists using this package feel that the airbrush effect, as a result of the i750 video processor's graphics capabilities, is more natural, with little or no lag or delay.

Time Arts also offers an optional library of over 40 antialias fonts for use in developing applications. Application developers can use these to design a unique look for an application, matching the fontstyle to the design.

Editing software

D/Vision from Touchvision Systems is a useful tool for editing motion-video sequences displayed using RTV. Using the compressed RTV files, video can be edited through a "point and edit" interface which allows you to build edited video sequences. The edited video sequences can then be played back and inserted directly into an application as RTV files.

V-Edit also offers an SMPTE (*Society of Motion Picture Technology Engineers*) time-code, frame-accurate readout. The SMPTE time code from the RTV edited files can be used for input into an on-line editing system to edit a final videotape prior to compression into PLV files.

Authoring software

Many application developers will elect to use an authoring environment instead of programming in their applications using the C language. The tradeoff of using authoring products is the ability to develop DVI® applications without the need for C programming expertise. The authoring products available allow access to most of the capabilities of the DVI® products. Some specialized capabilities are not available, and will require C programming. Several types of authoring products are available for the DVI® application development environment. They are different, and their capabilities should be judged carefully when you begin to plan which tool to use for your application development.

MEDIAscript

MEDIAscript from Network Technologies Corporation was the first DVI® application development software product developed to access the multimedia capabilities of DVI® products. It was developed by Arch Luther, who was part of the original development team when DVI® was invented at RCA's David Sarnoff Research Center in the mid-1980s.

MEDIAscript is based on a scripting language approach to developing software. Programs are written in simple English language text scripts, and provide access to most of the capabilities of the DVI® technology. Sophisticated applications can be developed very quickly using MEDIAscript. MEDIAscript also includes a selection of application development tools such as a bit-mapped editor, image processing tools, and memory management tools, to aid in application development. A run-time version of MEDIAscript is available for packaging and distribution with MEDIAscript-developed applications.

Authology:MultiMedia

Authology:MultiMedia from CEIT Systems, Inc. uses a point and click, character-based, window interface to allow a developer to create DVI® applications. Authology:MultiMedia includes a number of tools to control the presentation of different bit maps to create screens and menus. Either dynamic or bit-mapped fonts can be added to screens composited using Authology. Authology can then be used to string together multimedia elements into the program flow that makes the application. Using the point and click interface, the various multimedia elements are listed on-screen and can be sequenced in the application.

A run-time version of Authology:MultiMedia is available from Ceit Systems for distribution of applications.

DVI® DELIVERY SYSTEM

One of the most important factors when setting up your DVI® application development environment for a project is the specification for the delivery environment. Knowing the specifications for your target environment will allow you to make decisions to make sure your application will play back on your intended delivery systems. For example, some specialized applications in training and simulation may be designed to take advantage of DVI® technology's capability to support two monitors.

An example might be in a manufacturing process training application where the valve in a cooling system is graphically animated on the DVI® presentation monitor, while a schematic graphic of the overall system is rendered on the VGA display. The development environment for this application will be the two-monitor display system similar to the application development platform shown in Fig. 5.1, the more typical DVI® development system.

Base computer platform

The specification of a DVI® delivery system will many times be based on two factors: Does the application need to be installed into an existing base of computer platforms, and if not, what type of cost/system performance needs to be considered for this application? This last point must be considered when the delivery platform will be used for other applications. For example, a delivery system may be designed to access a text-based database in addition to stills and audio and video files in a DVI® application.

If low cost is a consideration, platforms based on Intel's economical Intel386 SX microprocessor may be the preferred choice. The Intel386 SX architecture provides adequate performance to access the capabilities of the ActionMedia system. In cases where more performance is desired for use with other applications, the Intel386 DX architecture, similar to the application development platform, running in a 25 MHz or 33 MHz mode may be desirable.

Your software design and other software needs will dictate the need for either two megabytes or four megabytes of system RAM on the delivery platform. For example, if the system will also be used to access Microsoft Windows™ applications, four megabytes of memory is desirable.

The selection of the hard drive in the system configuration is very important in the application design. Depending on the performance desired, an application may require some of the information stored and delivered on either a CD-ROM or LAN to be copied to the delivery system's hard drive for better performance. The hard drive will have to be large enough to temporarily store some of the DVI® application's files in addition to the other applications stored there.

If an existing computer system is being upgraded for the delivery of DVI® applications, a consideration during application development may be available hard-disk space for the application, or the flexibility of increasing the size of the hard-drive systems in the installed base of delivery hard drives. For example, if an installed base of IBM PS/2 delivery systems is equipped with 80-megabyte hard drives for typical office use, the decision may be required to either replace and upgrade the hard drive to allow multimedia information to be copied to a portion of the drive, or the application will have to be designed for accessing data either from a local area network or CD-ROM drive. The slower access time and data transfer rate of the CD-ROM drive will be a factor in many of the decisions in the application design.

The decision on how to distribute your multimedia applications must also be considered. As noted above, the CD-ROM or LAN are two options. If a small number of delivery systems will be dedicated to a multimedia application, the most cost-effective solution may be the use of streaming tape systems to load the multimedia information individually onto a number of delivery system's hard drives. Careful analysis of the delivery system needs will dictate the right media for transporting the multimedia information to the delivery system.

As our initial example in this section noted, there are also decisions to be made about the display monitors. Besides the decision between one or two monitors, a touch screen may be an option for a pointing device/input interface in place of a mouse. Touch screen kits or complete touch screen monitor systems are available for many popular VGA or NTSC display monitors. Your application needs will dictate your display system and input system needs.

DVI® delivery add-ons

Once your baseline delivery system is defined, you can add on the hardware components for DVI® application playback.

ActionMedia 750 delivery board

The ActionMedia 750 Delivery Board is available in two configurations, one with one megabyte of VRAM (*video RAM*), and one with two megabytes of VRAM. For most installations, the two-megabyte version of the delivery board is preferred. This is the most-often selected, and will allow your installed base of users to playback all DVI® applications.

Where economy is a consideration, and the user base of systems will be dedicated to specific applications, a one-megabyte ActionMedia 750 Delivery Board can save up to $500 per system. This is not without tradeoffs, however. The one-megabyte version of the board is limited when it comes to displaying large still images and for certain video transitions. For example, due to the VRAM space needed to display motion video, an application designed for playback from one-megabyte boards would have a delay or short period where the display screen goes black between video clips, while a two-megabyte board can hold a video image in memory to display while seeking to the new clip allowing a continuous display in the application design.

A dissolve transition between two nine-bit images illustrates another limitation. For a dissolve, both images have to be loaded into video memory. The dissolve shows one image, and transitions to the second one. At the higher resolutions, this may not be possible on a one-megabyte delivery board. Carefully evaluate the tradeoffs between one-megabyte and two-megabyte board configurations before finalizing on your delivery environment.

CD-ROM drive

Intel or IBM, or their designated resellers of ActionMedia products, can supply you with a list of the latest approved and tested CD-ROM players. If you are playing motion video from the CD-ROM, it is critical that you use a CD-ROM player that has been tested for continuous play.

Audio speakers

For speakers, there is a wide selection available to meet the needs of your application and budget. Self-amplified speakers are the most convenient, as they eliminate the need for a separate amplifier. These speakers were originally designed as an add-on to portable tape players, and are available in prices ranging from $25 to several hundred dollars a pair. The more expensive speaker systems will deliver high-fidelity audio, and may be inappropriate for office use. Inexpensive speakers can be excellent for training applications for the desktop. Another alternative for audio is a set of headphones.

The complete delivery system

Carefully evaluate your users' needs today, as well as the future. As you can see, even the simple selection of speakers can have an effect on production planning and design decisions. If your installed base will never need to use applications with high-fidelity sound, you shouldn't invest in high-fidelity speakers, or for that matter invest in a high-fidelity soundtrack during production. If there is no need for high-speed access to images for transitions, you may not want to specify a two-megabyte ActionMedia board in the delivery system. On the other hand, if your end users will be growing in their demand for multimedia features and capabilities, you must weigh the advantages of the highest quality delivery system against the trade-offs.

WHERE DO I FIND EVERYTHING I NEED?

This chapter outlines many of the systems and peripherals used for creating the multimedia elements which make up an application. They typically vary in price, based on quality. Many require specialized software as an interface to the ActionMedia application development environment. It is important to establish the system configuration for your application development, and to understand the capabilities and operation for your production. Where possible, running sample tests is the best way to learn how to use the equipment and to judge the potential quality and value for yourself.

Assembling all of the products desired for a DVI® application development system can be a daunting task. Several resellers are skilled in this area, and have the experience with DVI® applications to understand the compatible software and cable configurations for the different peripherals. The manufacturers of many of the peripheral products can also be of some help where they have experience with the DVI® products. Names and addresses of the manufacturers and DVI® ActionMedia VARs and resellers are listed in the resources section in Appendix B.

Sound complicated? Originally, so was the thought of using a personal computer for word processing, spreadsheets, and working with databases 10 years ago. Computers have gotten easier to use, while at the same time extending the capabilities of what we can do with them. The additional capabilities offered by the computer have required us to learn new tools to offer new capabilities. With these new tools and the understanding of how to use them, you can build exciting new applications which take advantage of the multimedia capabilities of DVI® products and the personal computer.

Chapter

6

Managing Multimedia Application Development

PROJECT MANAGEMENT FOR MULTIMEDIA

There are a number of different ways to think about the management of multimedia. You can look at each one from the perspective of a budget, a time line, the resources available, or the goals you have set out to accomplish. In reality, none of these perspectives are pure. Good project management turns out to be comparable to a juggling act. When the budget in one area falls short, there may be ways to use media production to enhance the end product; when the goals of the original video production team are not met adequately, there may be a way for the programmer to write a special piece of code to compensate for that fact. The good project manager needs to be able to ask questions that will help find solutions.

The good project manager is also a good people manager. Just as project budgets and goals must be balanced, the strengths of each team member must be identified, and must be accounted for in the overall development of the application. If a diverse team of people are coming together—as is often the case in multimedia development where film and video producers meet programmers for the first time—the project manager must define a working environment and protocol for the team members to work best.

In this chapter, we will examine a few tools that will help you to manage the logistics of the application development, including budget and time-line management. We will also give you some specific examples of a budget worksheet and management chart that we have used in our own application development. In addition, we will give you some ideas about how to communicate to your team and among team members during the application-development process. Managing a project for multimedia application development, like most projects, cannot be a formula, but these tools should help give you a framework to get projects off to a running start.

THE QUESTION OF BUDGET

If we ran a contest for the "most asked question about DVI® application development," it may well be, "How much does it cost to create DVI® applications?" This question is a bit like asking how much a car costs. You can purchase a car for less than $500—the question is, will that car meet your needs?

If you are working with a client or customer to develop an application, or if you are working internally at a company that is beginning to use multimedia products or create a multimedia product line, you will need to know generally how much you should expect to spend. A good set of questions to ask when evaluating the need for a multimedia solution follows:

What is the problem that needs to be solved?

What is the priority of the problem?

What is the scope of the problem in the eyes of the end user?

Is there a time frame that a solution is needed?

Is there a budget available to solve this problem?

Answers to these questions will bring a specific solution into focus. Finding the correct multimedia solution for a client, internal or external, is just good business consulting. The two most critical management tasks after determining that a multimedia application is the best solution are to create and manage the budget and to manage the project's time line.

Step number one—create a treatment

In order for a fair assessment of budget to begin, you must have some idea about the goals and the scope of the application. By creating a short, two-to-three-page document that outlines what the product is, you will force yourself, and your team, to answer these questions. This document is known as the *treatment*.

The first idea for a multimedia application may be sketched on a white board, on a scrap of graph paper, the margins of a Daytimer, or even on a word processor. But wherever they first appear, the idea makes it to paper somehow, as words, sketches, and symbols, or some combination of the above. Why paper? Few productions are produced entirely by one person. And even if they are, one person cannot remember all of the details of putting together a production.

The first formal document to appear as a part of the application-development process is usually the treatment. There is no formal format for a treatment that makes one application more effective than another in the end. The treatment must meet the needs of the team and the client, if appropriate. It is the document that first conveys the concepts that the team has agreed to be a part of the design of the application. It is the discussion of the team and the consensus of the team, written down for dissection, discussion, and most important, revision.

While there is no magic formula for the treatment document, effective treatments will contain some common information. These common elements call upon the three areas that are considered in the development of an application: the subject, the audience, and the setting. The treatment does not define the application, but usually goes far enough in the description of the application that a tone, communication goals, and a prototype walk-through of the users' experience are described.

Another valuable use for the treatment is as a part of a product proposal. As you read over the sample treatment here, it sounds very much like a proposal that you may give to a client, or to a superior who needs to approve your product idea. (A full-blown proposal will probably also include a preliminary time line and a budget.) The first treatment is used for discussion, and will probably need revision. The treatment that is finally agreed upon by the team is usually a milestone in the project path. In some cases, a proposal can be charged to the client, especially if the creative team is well-known or has worked with the client previously (and with success). Many times the proposal is developed at the producer's own cost.

The treatment sets the stage for the next steps of development: preliminary budgeting, project planning, and creating flowcharts, storyboards, and scripts. Some product proposals will have storyboards and flowcharts attached; but at the least, a preliminary budget and time line are prepared.

Creating and managing a budget

The treatment, along with communicating the application as a whole, helps to determine what the application may look like and feel like, and can give the project manager (or multimedia producer) a sense of

SAMPLE APPLICATION TREATMENT

LOCAL AREA NETWORK (LAN) TRAINING APPLICATION TREATMENT

Materials

The material to be used for this DVI® application is Anderson Soft Teach's video course on local area networks. This course is a general overview of networks, how they operate, and what advantages they offer to the user. We will use the original videotape, along with specially created graphics, stills, and additional voice-overs (using the same talent from the videotape), and create an interactive program. DVI® technology will be used as the development platform, and for the delivery of this application.

Approach

This application will cover information to introduce users to local area networks. The following content will be covered:

- Definition of local area networks
- Configuration of local area networks
- Explanation of fileserver function
- Three types of networks
- Advantages of a network
- Sharing data and application programs
- Record and file locking
- Security features

Using CEIT System's Authology:MultiMedia authoring package, we will create graphics and animation to illustrate major concepts. In addition, "hot buttons" will allow the user to seek more information in any topic area by selecting certain terms.

Design

The application will begin with a title screen that will move the user to an introduction segment that describes how to use the program. A menu listing will then appear. The two general topics will be, "How to Use a Local Area Network," and "What Are Local Area Networks?".

"How to Use a Local Area Network" will cover topics and terms such as application programs, data files, printer, and modem. An index screen will also be available for a quick search on any term. Customized instruction, such as the segment that explains the specific use of a printer, will be able to accept information about the user's specific configuration. This may be used to show a map of where printers in a building are located, for example.

"What Are Local Area Networks?" will cover general definitions of networks and their components. Terms like network, fileserver, network-operating system, printer, plotter, and modem will be covered in video, audio, graphics, and by text definitions, as well. A quiz and suggested remediation path will accompany this part of the application.

Production Company Name
Title: "Learning the Ropes"
Project: Sample Training Program

Last Revised:	4/1/91		Client:	XYZ Corporation
Completed by:	SRM		Producer	J. Smith

	Quantity	Time/Units	Rate	Total	Notes/Comments	
Production Staff						
Executive Producer	3	Week	3000	9000	Contract	6750
Producer	35	Week	2400	84000		
Associate Producer	0	Week	2000	0		
Production Manager	35	Week	1400	49000	Contract	36750
Production Assistant	0	Week	1000	0		
Controller	6	Daily	250	1500		
Administrator	4	Daily	150	600		
SUB				$144,100	Save	14500
Creative Staff						
Scriptwriter	9	Daily	1500	13500		
Graphic Design	12	Daily	450	5400		
Interactive Designer	5	Daily	1500	7500		
Instructional Design	5	Daily	1000	5000		
Programming Design	0	Daily	1200	0		
Researcher	15	Daily	300	4500		
SUB				$35,900		
Production						
Art Prod.		Contract				
Computer Artists	400	Hour	80	32000	Contract	24000
Stills Photography		Contract	3000			
Scanning	500	Unit	18	9000		
Soundtrack		Contract		0		
Voiceover Talent	1500	Contract		300		
Studio		Contract	2000	200		
Music License Fees		by Cut		0		
Editor	16	Hourly	100	1600		
Composer		Contract		1000		
Sound EFX	0	by Cut		0		
Video Production		Contract				
Video Producer	0	Daily	0	0		
Director	0	Daily	0	0		
Assistant Director	0	Daily	0	0		
Still Photographer	0	Daily	0	0		
Cameraman	0	Daily	0	0		
Lighting Dir.	0	Daily	0	0		
Gaffer	0	Daily	0	0		
Gripp	0	Daily	0	0		
Sound	0	Daily	0	0		
Production Assistant	0	Daily	0	0		
Talent	0	Daily	0	0		
Makeup	0	Daily	0	0		
Script Coordinator	0	Daily	0	0		

Figure 6.1 Multimedia Production Budget Showing All Line Items.

	Quantity	Time/Units	Rate	Total	Notes/Comments	
Video Shoot Equip.	0	Daily	0	0		
Tape Stock	0	By Reel	0	0		
Sets and Props	0	Contract	0	0		
Teleprompter	0	Daily	0	0		
Production Travel/Ship		Total		0		
Video Post Production	16	Hourly	350	5600		
Video Compression	20	Minute	250	5000		
Media Costs & Racking	1	Fixed		170		
SUB				$54,870	Save	8000
Programming						
C Programming	0	Daily	0	0	Contract	0
Author	45	Daily	400	18000	Contract	13500
Prototyping		Total				
Application Testing	0	Daily	100	0		
Programming Equip.	0	Daily	280	0		
SUB				$18,000	Save	4500
Duplication						
CD Prep		Contract		120		
CD Pressing		Contract		1500		
CD-ROM Disks		Contract		750		
Packaging Design		Contract		300		
Packaging Prod.		Contract		0		
SUB				$2,670	Save with Contracts	19065
Grand Subtotal				$255,540		$236,475
Contingency	15%			$38,331		$35,471
Total				$293,871		$271,946

Figure 6.1 *(cont.)* Multimedia Production Budget Showing All Line Items.

what budget items will be needed to be included. The format we use for budgeting is a Microsoft Excel spreadsheet template working under Windows operating environment. A copy of a typical budget is shown here as Fig. 6.1. This budget represents the multimedia presentation presented in Chap. 3.

The costs represented in Fig. 6.1 are calculated based on real projects we have completed in the past. These numbers are likely to change if you are subcontracting work out, or doing it in-house, and will change according to geographic regions. The costs here represent two fairly expensive areas, the East Coast and West Coast metropolitan areas. The major categories of cost are personnel, media elements production and acquisition, overhead, contingency, and profit fees.

You recall from Chap. 3 that this presentation used no original photography or motion video. Original music production and origi-

Production Company Name
Title: "Learning the Ropes"
Project: Sample Training Program

Last Revised:	4/1/91	Client:	XYZ Corporation
Completed by:	SRM	Producer	J. Smith

	Quantity	Time/Units	Rate	Total	Notes/Comments
Production Staff					
Executive Producer	3	Week	3000	9000	Contract 6750
Producer	35	Week	2400	84000	
Associate Producer	0	Week	2000	0	
Production Manager	35	Week	1400	49000	Contract 36750
Production Assistant	0	Week	1000	0	
Controller	6	Daily	250	1500	
Administrator	4	Daily	150	600	
SUB				$144,100	Save 14500
Creative Staff					
Scriptwriter	13	Daily	1500	19500	
Graphic Design	12	Daily	450	5400	
Interactive Designer	5	Daily	1500	7500	
Instructional Design	5	Daily	1000	5000	
Programming Design	0	Daily	1200	0	
Researcher	15	Daily	300	4500	
SUB				$41,900	
Production					
Art Prod.		Contract			
Computer Artists	400	Hour	80	32000	Contract 24000
Stills Photography		Contract	3000		
Scanning	500	Unit	18	9000	
Soundtrack		Contract		0	
Voiceover Talent	1500	Contract		300	
Studio	1	Contract	2000	2000	
Music License Fees		by Cut		0	
Editor	26	Hourly	100	2600	
Composer		Contract		1000	
Sound EFX	0	by Cut		0	
Video Production		Contract			
Video Producer	5	Daily	300	900	
Director	3	Daily	800	2400	
Assistant Director	2	Daily	500	1000	
Still Photographer	0	Daily	0	0	
Cameraman	1	Daily	800	800	
Lighting Dir.	2	Daily	500	1000	
Gaffer	2	Daily	300	600	
Gripp	1	Daily	300	300	
Sound	1	Daily	300	300	
Production Assistant	0	Daily	0	0	
Talent	0	Daily	0	0	
Makeup	0	Daily	0	0	
Script Coordinator	0	Daily	0	0	

Figure 6.2 Production Budget with the Addition of 10 Minutes of Motion-video Production.

	Quantity	Time/Units	Rate	Total	Notes/Comments	
Video Shoot Equip.	1	Daily	1500	1500		
Tape Stock	5	By Reel	100	500		
Sets and Props	0	Contract	0	0		
Teleprompter	0	Daily	0	0		
Production Travel/Ship		Total		0		
Video Post Production	26	Hourly	350	9100		
Video Compression	30	Minute	250	7500		
Media Costs & Racking	1	Fixed		170		
SUB				$79,970	Save	8000
Programming						
C Programmer	0	Daily	0	0	Contract	0
Author	45	Daily	400	18,000	Contract	13,500
Prototyping		Total				
Application Testing	0	Daily	100	0		
Programming Equip.	0	Daily	280	0		
SUB				$18,000	Save	4500
Duplication						
CD Prep		Contract		120		
CD Pressing		Contract		1500		
CD-ROM Disks		Contract		750		
Packaging Design		Contract		300		
Packaging Prod.		Contract		0		
SUB				$2,670	Save with Contracts	27,000
Grand Subtotal				$286,640		$259,640
Contingency	15%			$42,996		$38,946
Total				$329,636		$298,586

Figure 6.2 *(cont.)* Production Budget with the Addition of 10 Minutes of Motion-video Production.

nal artwork were prepared for this project. Most of the work on the presentation was subcontracted, not done in-house. Since the production represented fairly steady work for several of the team members, contracts were negotiated at 75 percent of the normal hourly rate. (These are noted on the outermost column on the spreadsheet, and used in the bottom-line calculations.) This is a fairly common practice when more than a few days of time on any given production is required. It helps reduce costs for the production, and helps the individuals plan their work (and income) more effectively.

All programming was done in Authology:MultiMedia on this presentation. It was very straightforward, since it was a linear presentation, so costs in this area are light compared to another, more complicated, programming effort.

Original motion video production

One area that can change overall costs of a production is original motion video production. If we had added 10 minutes of motion video original production to this budget, the bottom line would change as illustrated in Fig. 6.2.

The change in the bottom line of this budget is due to the following line item changes:

Video producer—five days

Scriptwriter—four additional days

Crew—one day

Edit suite—one day

Video compression—10 minutes at $250 a minute, plus media and racking fees

Note that the contingency changes as a result of these additional costs.

This template can be used to calculate and to discuss with a client or manager approximate costs of different types of applications. It is only a guideline, however. If we look at the video production area, for example, the charges and expenses can vary greatly. Industrial production is likely to be an order of magnitude less than a production by a Hollywood video production studio. Production that requires actors, lighting directors, or special-effects editing will be much more expensive than productions without these requirements.

Other interesting factors to consider when doing motion video are the hidden costs to the client. For example, for one production we were interested in shooting original footage in the Intel factory that manufactures Intel486™ Microprocessor (i486) components. We determined that the cost of disrupting the production line and potentially introducing contaminants in the production area was too costly, and opted to use stock footage and stills to illustrate the concepts. Often in this type of situation, a designer will elect to create a graphical simulation or use animation to illustrate a point that cannot be efficiently videotaped.

Budget implications of authoring vs. programming

The spreadsheet quickly identifies the area of most cost, and can help to find areas that are too costly for the payback. The original footage example above is one. Another may be programming costs. The multimedia presentation used as our model was created using Authology:

	Quantity	Time/Units	Rate	Total	Notes/Comments	
Programming						
C Programmer	50	Daily	600	30,000	Contract	
Author	0	Daily	400	0	Contract	25,000
Prototyping		Total				
Application Testing	0	Daily	100	0		
Programming Equip.	0	Daily	280	0		
SUB				$30,000	Save	5,000

Figure 6.3 Cost Impact of Programming instead of Authoring.

MultiMedia. What would the costs have been had it been programmed in the C language?

The original budget called for forty-five days of authoring with Authology:MultiMedia. In this particular instance, we can assume that the programmer is familiar with DVI® Technology and the ActionMedia 750 Software Library. Since this is a multimedia presentation with very limited interactivity, the number of days to program stays the same. The cost of a programmer, however, is a bit more than an author, as shown in Fig. 6.3. The total cost for this project does not change a lot. But over the course of a longer, more complicated production, the costs would change significantly. You will need to consult a programmer to come up with more accurate numbers.

Remember that programming is a creative process, and that changes will most likely occur in the design of a product as a result of feedback from the programmer and other people on the creative team. Some programmers will adjust their time and cost estimates with this in mind; other programmers will give you a "bare-bones" estimate of their time without building in any cushion. Before you commit to a budget, it is important to know how the programmer came up with any estimates. If a cushion was not accounted for in the original estimate, it would be wise to add one.

Using a spreadsheet in design

The advantage of working in a computer environment for budgeting work is that you can try options on, and determine the impact on the budget almost instantly. For instance, the dollars added to this budget by programming in C can now either be accepted or not. Alternatives can be considered and figured without too much effort. Typically, the first budget spreadsheet is only a working model that is massaged and changed while juggling client expectations, costs, and the available budget.

Our choice for this product was to stay with authoring as our method of implementing the application. This is not to say that authoring is always the best choice for all applications. The presentation, which would not call on any specific functionality of C, would not have benefitted.

Budgeting for a large-scale product

The presentation we have used as an example here, however, was not a full-blown application. Many of the anticipated line items in the budget lines remain blank. A budget for a large-scale application will look very different and, as you may guess, will be more costly. Each product will have a custom budget, according to the content, the complexity of the programming, and the need for original artwork, music, or video production.

We have included one example of a full-blown application, shown as Fig. 6.4. This budget is based on the real figures, for an actual training application. The application is a training tool for a retail sales organization that has a high turnover, and employs a lot of teenage youth. The training application is instructional, but also has a high entertainment value and a game component for each topic. The game component was designed to motivate young employees to use the application regularly.

The application is first in a series that would use the same format, some of the same characters, and often would share programming code, graphic art design, and content expertise. There would be between 15 and 20 "sister" applications, to be produced over a period of two years.

It was determined that the design of this initial application would be critical, and would be costly. But this was justified as a very necessary expense, since a major goal was to engage teenage employees in instruction and game play. The design for this application will carry through to the subsequent applications, and is actually the design of an entire course, not just one module.

Likewise, the programming and authoring will have a high cost initially, but will come down for the subsequent applications as the tools and the code are created. The software production tools and code created for this application will be used in the same format for the additional training applications.

There will also be some savings if this project is done on a contract basis for some key players—like the producers, artists, programmer, and author. These professionals will significantly cut their costs if promised work for a period of two years. This savings is realistic (about 25 percent), given both the duration and scope of the project as a whole.

The budget below reflects costs for the first training application. The total reduction in cost for the additional applications was project-

Production Company Name
Title: "Retail—Knowing Your Customer"
Project: Sample Training Program

Last Rev.:	4/1/91		Client:	XYZ Department Stores	
Completed by:	SRM		Producer	J. Smith	

	Quantity	Time/Units	Rate	Total	Notes/Comments	
Production Staff						
Executive Producer	4	Week	3000	12000	Contract	9000
Producer	10	Week	2400	24000		
Associate Producer	9	Week	2000	18000		
Production Manager	6	Week	1400	8400	Contract	6300
Production Assistant	10	Week	1000	10000		
Controller	10	Daily	250	2500		
Administrator	8	Daily	150	1200		
SUB				$64,100	Save	5100
Creative Staff						
Scriptwriter	18	Daily	1500	27000	Contract	20000
Graphic Design	20	Daily	450	9000	Contract	8000
Interactive Designer	15	Daily	1500	22500	Contract	18000
Instructional Design	15	Daily	1000	15000		
Programming Design	5	Daily	1200	6000	Contract	5000
Researcher	0	Daily	300	0		
SUB				$79,500	Save	12000
Production						
Art Prod.		Contract				
Computer Artists	400	Hour	80	32000	Contract	24000
Stills Photography	4		1500	6000		
Scanning	500	Unit	18	9000		
Soundtrack		Contract		0		
Voiceover Talent	2	Contract	1500	3000		
Studio	2	Contract	2000	4000		
Music License Fees	2	by Cut	300	600		
Editor	16	Hourly	100	1600		
Composer		Contract		0		
Sound EFX	0	by Cut		0		
Video Production		Contract				
Video Producer	9	Daily	300	2700		
Director	8	Daily	800	6400		
Assistant Director	6	Daily	500	3000		
Still Photographer	0	Daily	0	0		
Cameraman	3	Daily	500	1500		
Lighting Dir.	4	Daily	400	1600		
Gaffer	4	Daily	300	1200		
Gripp	4	Daily	300	1200		
Sound	3	Daily	300	900		
Production Assistant	6	Daily	250	1500		
Talent	3	Daily	1000	3000		
Makeup	3	Daily	300	900		
Script Coordinator	3	Daily	250	750		
Video Shoot Equip.	3	Daily	2500	7500		

Figure 6.4 Production Budget for Large-scale Multimedia Project.

	Quantity	Time/Units	Rate	Total	Notes/Comments	
Tape Stock	25	By Reel	50	1250		
Sets and Props	1	Contract	5000	5000		
Teleprompter	3	Daily	300	900		
Production Travel/Ship	1	Total	500	500		
Video Post Production	25	Hourly	350	8750		
Video Compression	15	Minute	250	3750		
Media Costs & Racking	1	Fixed		170		
SUB				$102,670	Save	8000
Programming						
C Programmer	0	Daily	0	0	Contract	0
Author	14	Daily	400	5600	Contract	4200
Prototyping	0	0	0			
Application Testing	5	Daily	100	500		
Programming Equip.	18	Daily	280	5040		
SUB				$6,100		
Duplication						
CD Prep	1	Contract	1500	1500		
CD Pressing		Contract		1500		
CD-ROM Disks		Contract		750		
Packaging Design	1	Contract	500	500		
Packaging Prod.	1	Contract	8000	8000		
SUB				$22,890	Save	1400
					Save with Contracts	26500
Grand Subtotal				$275,260		$248,760
Contingency	15%			$41,289		$37,314
Total				$316,549		$286,074

Figure 6.4 *(cont.)* Production Budget for Large-scale Multimedia Project.

ed to be about 40 percent, based on consulting with industry professionals who have done this type of coursework development before.

PROJECT TIME LINE

A project time line and a budget often go hand-in-hand. A time line is a second blueprint for any multimedia application producer. There are many critical steps in the development of a multimedia application. If the work is being done by a team, the producer must make sure that milestones are met. Content must be produced and in the programmer's hands for a certain amount of time for success. Script must be delivered in time for the production to take place. The list goes on and on.

The project time line is another sanity check for the team before work begins, as well. If a product must be completed for shipment to a

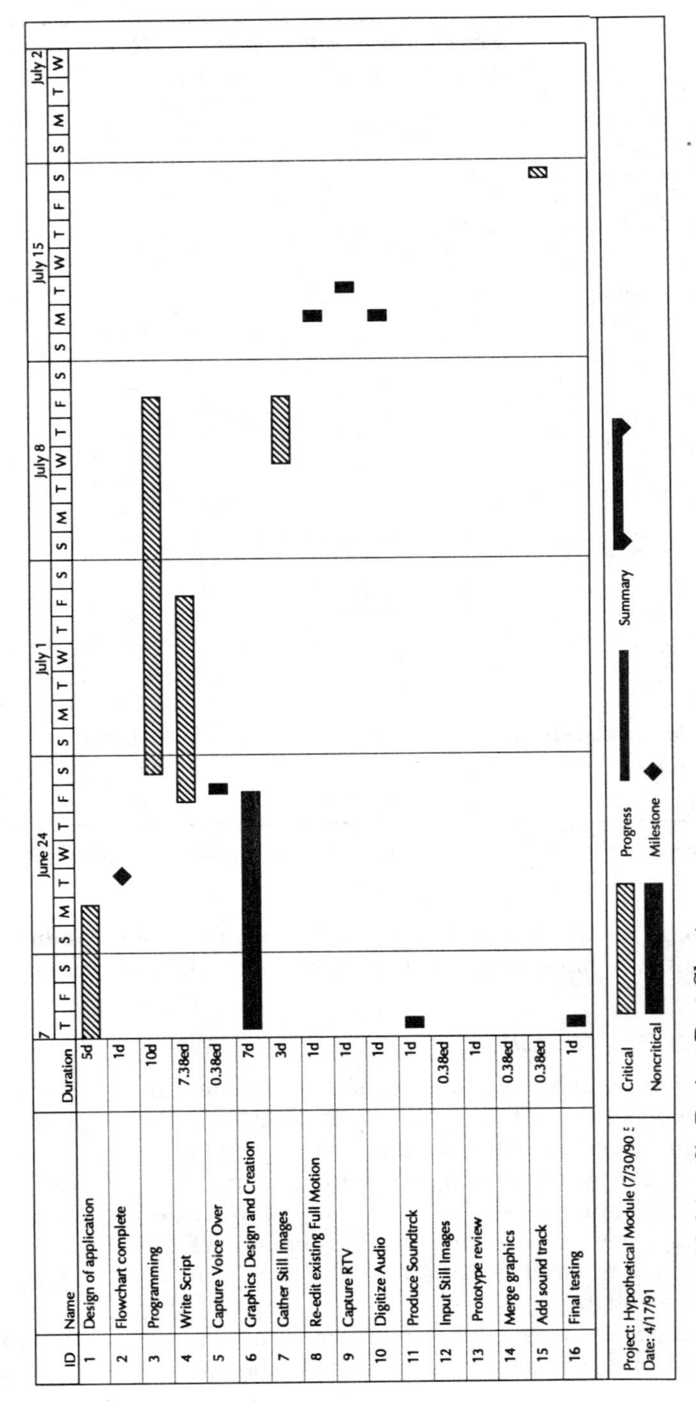

Figure 6.5 Multimedia Project Pert Chart.

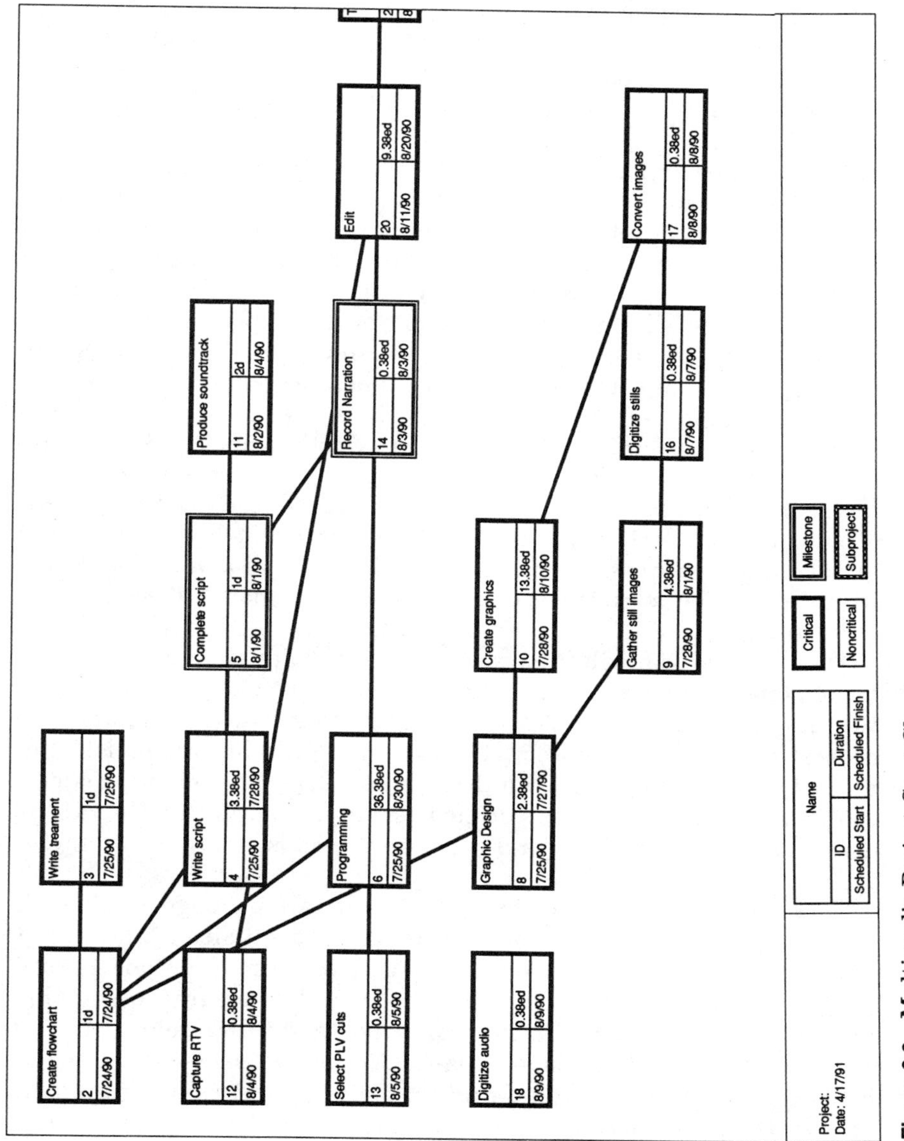

Figure 6.6 Multimedia Project Gant Chart.

packaging firm for Christmas delivery, for example, the producer and other team members can "back-time" from that critical date to determine whether the product is feasible.

The tool we have used to plot our projects is Microsoft Project for Windows. This software allows you to use your Windows-equipped personal computer for budgeting (with Excel) and project planning. Later, we will show you tools for Windows for creating flowcharts and scripts. Having all of this information in one place, and in common formats, will make your life as a producer and team member much easier.

Project plans created from this software can be either Pert charts or Gant charts. The project plans for the multimedia presentation we have been using as an example are shown in Figs. 6.5 and 6.6. The important milestones and critical path to complete the presentation are identified right on the plan. Note that on these plans, the design phase is minimal, while the production, including original artwork production, and programming phases are longer. In an interactive application, or one that has a broad and deep treatment of subject matter, the design phase will be much longer, and will involve more people. For instance, the interactive training example we used earlier to review budget had a three-month design phase built into it.

The way that you will use project planning software is really a matter of personal choice. We have found that this software helps to identify how and when different project activities overlap, and work in concert toward critical dates. The same thing can be accomplished in a linear plan created on a word processor. The real advantages of a tool like this one are that critical schedule adjustments can be assessed almost instantly, and it provides feedback and accountability to project team members.

Task paths can be defined using this software package, or people's roles as paths can be defined. Still-image acquisition and script writing may be tasks, for example. Yet, there may be one person who is responsible for both of these tasks. For a large team, this provides a big picture for everyone. For a smaller team, critical dates and workload can be evaluated. You may uncover the need to hire a script writer when you find that your photographer, designer, and writer are one and the same person—with critical path items due all in the same week.

Task paths from Project can also be used as objectives by the employee. This management tool allows the producer to specifically track and support activities that have been identified as critical for success of the project, and to meet regularly with team members to understand progress, problems, and solutions.

Working with a client

If you are creating a multimedia application for a client, project planning software provides an excellent planning vehicle for the client and you to work together. A client who is familiar with project planning tools, and with managing with these types of tools, will appreciate the framework.

The client who is participating in the production of multimedia for the first time will see the complexities of the production, and will appreciate the need to have script, treatment, and storyboards reviewed and approved in a timely way. In the rare instance that this does not occur, the impact can be assessed and delivered to the client in the project-planning format.

Changing the product

The flexibility of this tool, like preparing the budget in Excel, encourages you to try out changes and alternatives in your production. If we added 10 minutes of original motion video to our presentation, as we did when going over the budget, where would it impact the time line? We know that it would have to come in somewhere after script writing and before final integration of all the elements. If the production is added, we can easily see that we either need to create another parallel task track that is staffed by others, or will need to push back the final integration, editing, testing, and delivery of the application. These changes are illustrated in Fig. 6.7.

Microsoft Project is a very complex tool that allows all types of interactions of the data around a project, including rates and cost of personnel, for example. Only the time line planning portion of this tool has been examined here.

The initial project time line is like the preliminary budget, and should be used in the same way—as a planning guide. Once the application product is underway, this same tool can be used to create detailed product-development time lines and paths. Each team member can create one for her or his area, if needed. These tools are excellent for communication, organization, and for tracking progress against goals.

THE PRODUCT PROPOSAL: TREATMENT, BUDGET, AND TIME LINE

These three elements of a product proposal are an exciting first step on your way to producing and managing multimedia application development. They may not feel very creative, and may not hold the interest of the team that will eventually make the personal computer

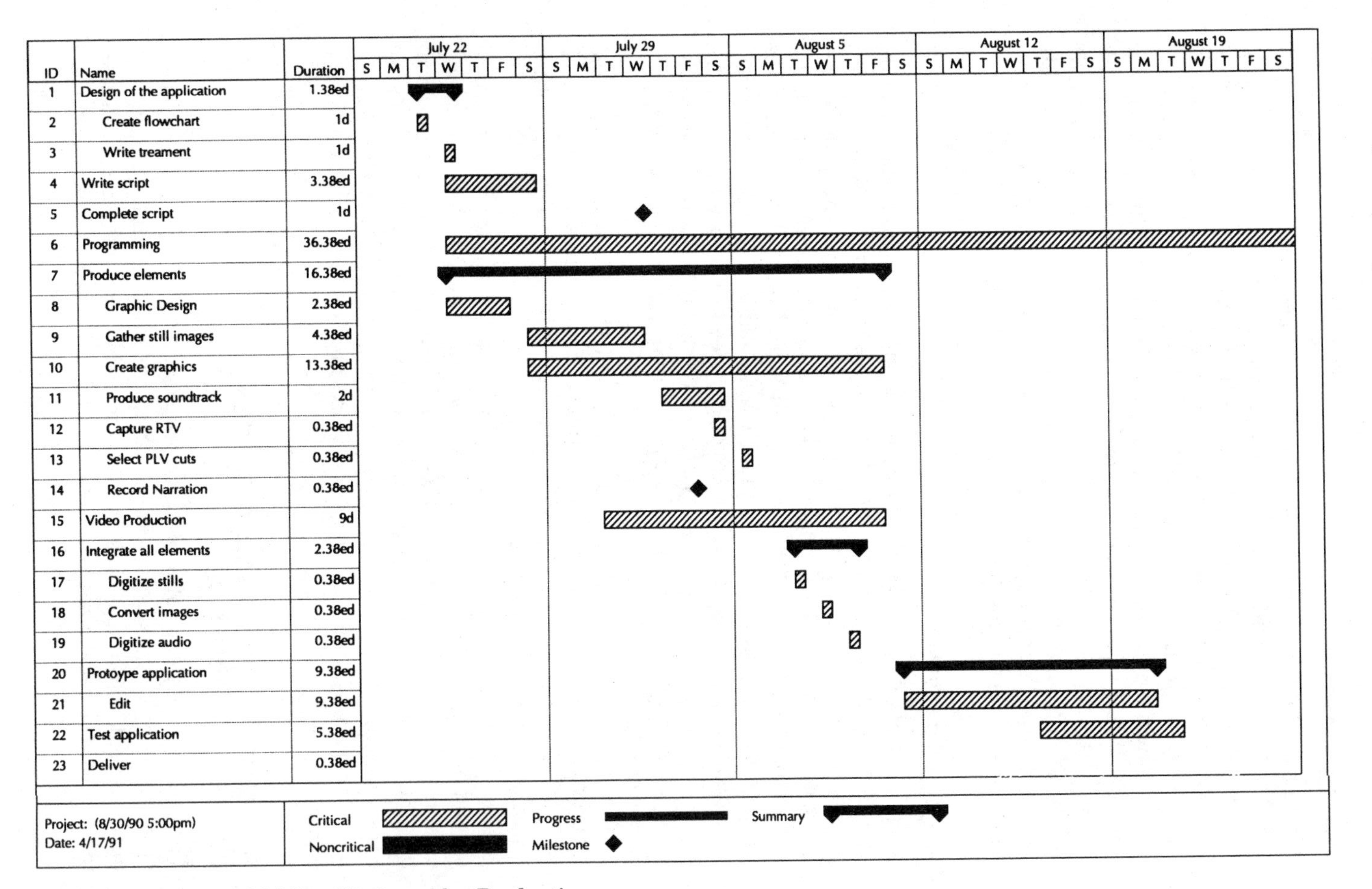

Figure 6.7 Impact of Adding Motion-video Production.

look and sound like it will with multimedia. These activities are not just a necessary evil, however. When you have to create a budget and a time line, and when you get others to commit to working with you in the development of these blueprints, you are taking the first steps in building a creative team.

We have found that artists and programmers appreciate this process, when made a part of it. Many creative ideas can come from asking the question, "How could we accomplish this same goal differently?" The DVI® application-development environment has many alternatives—motion video can be mixed with new audio as it was for our multimedia presentation example; animation can be created in a paint package or by writing a routine in C or an authoring package.

Many of these capabilities will be covered in Section 2 of this book. You will discover others as you work with DVI® products more. But most exciting of all, perhaps, is the fact that you and other developers will actually invent techniques as you gain experience and as you search for creative ways to overcome the challenges of multimedia application development. By involving the creative team up-front with the challenges faced in production planning, these new possibilities will come to the forefront early in your product planning.

Chapter

7

The Blueprints of Production: Flowcharts, Storyboards, and Scripts

If you have developed software for a personal computer before, whether for entertainment, computer-based training, or for interactive videodisc, you are probably familiar with the concepts we will cover in this chapter. The creation of the flowchart, storyboards, and scripts for any interactive, multimedia application is an integral part of the design process. This step of development lays the groundwork for how the content will be presented.

Whenever you talk with a programmer of a multimedia application, you are likely to hear that the programming is the most important part of application development. If you talk with a graphic artist, you will hear how the graphic design and overall look of the application is the most important element. Video producers, content experts, and interactive designers will all attest to the importance of the expertise they provide. The truth is that each is equally important. It is critical that the team of people, or the collective talents represented, agree on the goals, content, audience, and approach they are taking with the development of the application.

Whenever we have presented the information that follows in this chapter at conferences and workshops, there has always been healthy debate about the format and methods we propose. At times, there has been serious question about the need for this type of documentation at all. There is strong temptation to dive into the development of an application immediately after its conception and funding. After all, there are tools available that will allow you to prototype an application with the DVI® workstation fairly quickly.

The DVI® workstation gives you the ability to create sample screens, and to link them together to show a prospective manager or client how your concepts will work. You can collect and assemble video stills, audio, and motion video right at your desktop. This approach works very well for demonstrations or prototypes. However, the deeper, more complete your multimedia application is, the more you will benefit from early documentation.

You would never dream of shooting a movie or television program without a detailed production plan. Likewise, a software developer needs to complete a technical specification or architecture for any piece of software planned. These documents are your production plan and technical specification.

DESIGN DOCUMENTS

The three documents we propose are the flowchart, storyboards, and scripts. A flowchart is a visual blueprint of how the interactive program works. The flowchart is typically used by the interactive designer and programmer to determine how and when each media element will appear to the person using the application. Storyboards are rough drawings of how the screen will appear. They are used by all the team members, including the graphic artist, interactive designer, video producer, and the programmer. A storyboard and flowchart are very much related, and in some cases, can be one and the same. Scripts are usually used by the production team. Scripts are also used by the programmer and graphic artist to understand more about the tone and flow of the application.

Each of these documents is an important team-communication tool. These documents verify that all of the team members are working toward the same goals. They are the best way to find difficulties in programming, graphic art design, video production, or interactive design before they become entrenched in the application.

If a button must be available on the screen at all times, for instance, the graphic artist will know that from the storyboard, and will be able to design original artwork to include it. If there is no escape from a certain path in an application, the flowchart will identify that mistake to the interactive designer and the programmer. And if the script has two-to-three-minute soliloquies, the video producer, application product manager, interactive designer, and script writer can correct it in a script review session early in the development process.

These documents are also valuable in the last stage of application development, when documentation must be written prior to publication. If the application is well-documented during its development, the task at the end is all the simpler.

The order in which these documents are created is probably not important. In fact, if more than one person is involved in creating the product, it can be created as a team effort. Since multimedia is such a visual medium, storyboards may be the best place to begin.

Storyboards

A storyboard is a visual representation of your application. Storyboards are almost always used in the creation of video. You may be familiar with their use in advertising campaigns. In print media, the analogous document may be the draft layout of a newsletter or flyer. You can use storyboards to represent still frames, full-motion video, or graphics.

Storyboards can start as simply as pencil sketches with stick figures. This first direction can be handed over to an artist or designer to illustrate your first ideas. These first sketches can be elaborated by the artist on paper, or on a DVI® system. When you are the sole production staff, storyboards are the vehicle for you to sketch out your ideas and concepts, and to start playing "what if" to the visual content and flow.

The first storyboards can also be a part of the treatment, as discussed in Chap. 6. They help set the tone, and define the flavor of the application. Will your application present clean, uncluttered choices, or an invitation to roam and explore information? Are you trying to convey a certain corporate image? Should a logo appear on every screen? Who are the characters in the application? What are the settings?

Storyboards are a way to put the team's creative ideas into a visual presentation. They will progress in stages parallel to the development of the project. Early storyboards may only show full-screen images. Later versions will need some text or icons to explain the interactions available to the user. The storyboards will also reference the script. Storyboards may also be used as presentations to clients or management, or may be kept internal to the development team.

If you will be presenting storyboards as part of a management or client review, you may want to use a software package like Micrographx or Power Point (both are available for Windows) to create a sample screens and print them for distribution. Be aware that when you use a tool like this, much of what you are trying to convey

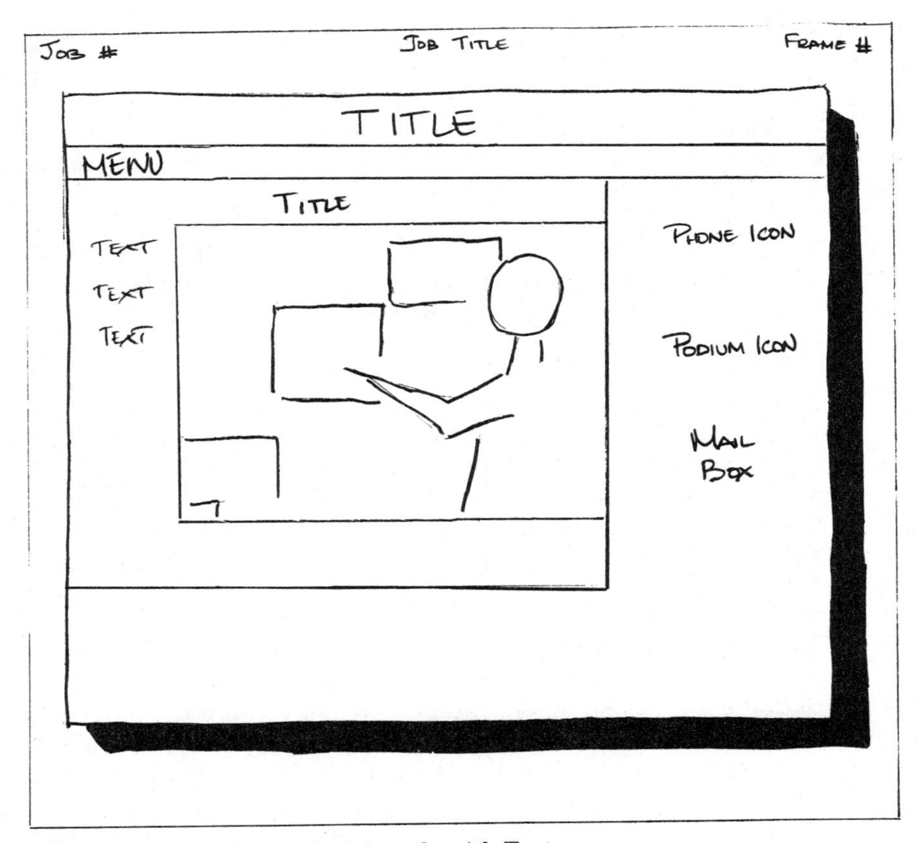

Figure 7.1 Stick Figure Storyboards with Text.

is lost because you are working with the limitations of these tools. Images will not have the depth and richness of DVI® displays, and no audio is available. For some managers and clients, especially those who have experience with multimedia or video production, this is not a problem. For others, you may find that they have a difficult time making the leap between what you are showing them and your conception of a final product. You may want to actually create the screens on the DVI® workstation, and print them to a video printer or even to a laser printer to help bridge this gap.

Some examples of storyboards are shown in Fig. 7.1, a stick-figures-and-text sketch that may be used internally as a planning document; Figure 7.2, a more detailed storyboard based on photocopies of actual images that may be used in the application; and Fig. 7.3, a complete storyboard, created by scanning images into the computer and creating computer graphics that can then be printed.

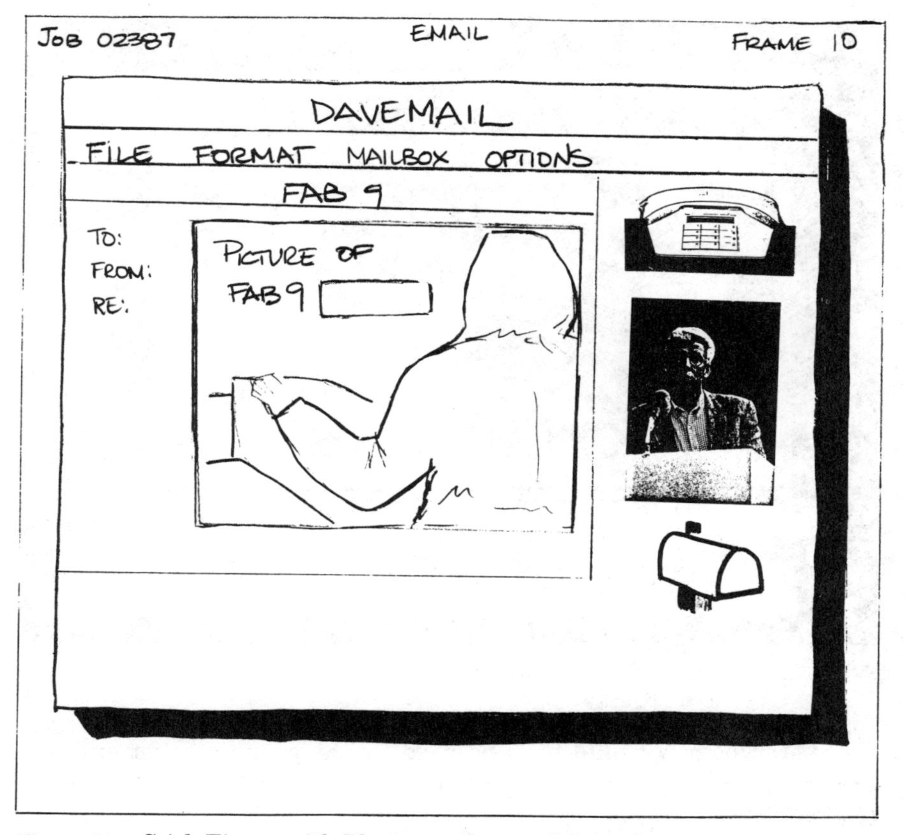

Figure 7.2 Stick Figure with Photocopy Images Inserted.

We recommend right from the beginning that you assign file names to the different multimedia elements for organization and planning purposes. As soon as you acquire an element, it can be cataloged, and the file name placed on the storyboard. These file names will correspond to file names in the flowchart, too, as you will see in later examples. That way, all team members can be consistent about how they are identifying images or elements.

A storyboard usually does not show every screen or element, but shows major categories of screens and how they will work. If you are creating a multimedia encyclopedia, or parts catalog, for example, you may create a screen that shows how a text search works. Each text search would work exactly the same as any other one. You may also show how text would be highlighted to designate that there is a still associated with it. The placement of any still, and how audio is accessed, will probably only have to be documented one time. Each of

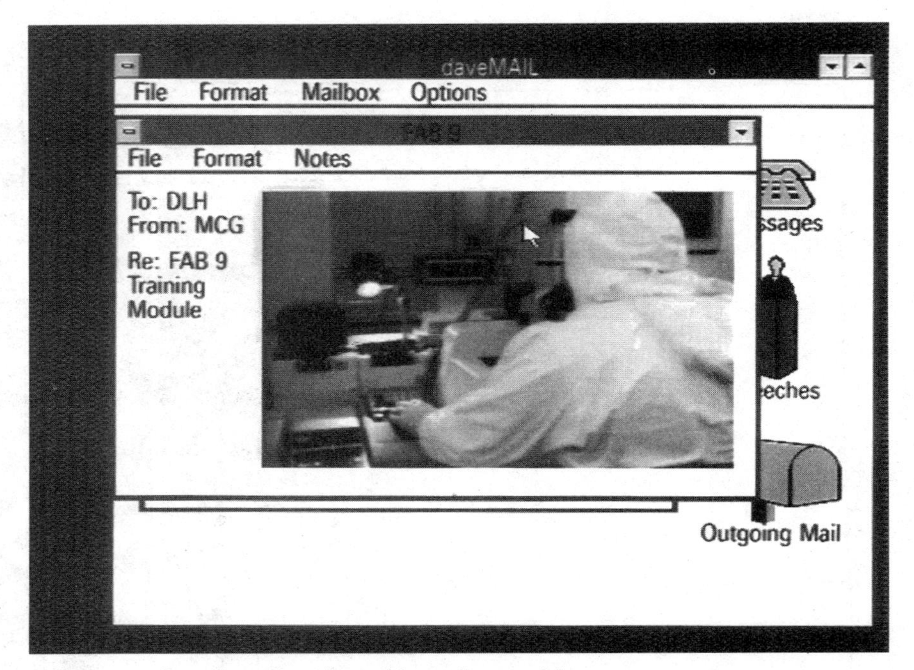

Figure 7.3 Complete Storyboard from Scanned Images.

these storyboards would then be used by the graphic artist and programmer to create a consistent interface to all the same types of interactions.

Flowcharts

Flowcharts are to interactive design what outlines are to writing. Flowcharts contain the character of the application, and can be read much like a blueprint to a home or building. If you are not the programmer, a flowchart provides you with important information. It visually describes the complexities of the application's software program flow, and the relationships between the content and the end user. As you gain experience reading a flowchart, you will be able to understand the flow of an application very quickly, merely by looking at the flowchart.

The flowchart for the multimedia presentation described in Chap. 3 is linear. While the media elements vary, the flow of the presentation has a clear line from start to finish. A flowchart is probably not even needed for a presentation like this one, except to keep all of the visual elements and audio files cataloged. Here, we have included an

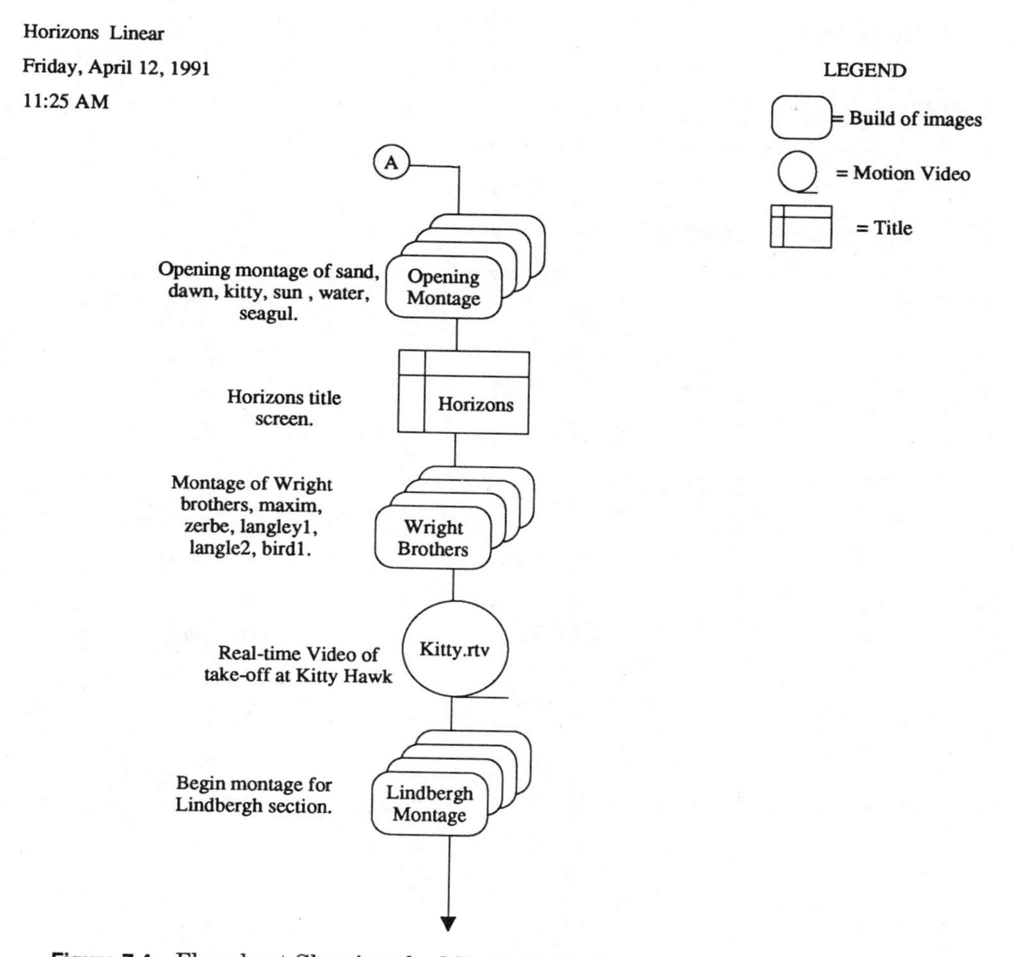

Figure 7.4 Flowchart Showing the Mix of Media Types.

edited version of the flowchart for your information. Note that in our example, different media types are designated by different shapes (see Fig. 7.4).

This presentation may be changed, however, so that it is interactive. The following flowchart, Fig. 7.5, illustrates how this same presentation could be organized differently. The introduction would stay basically the same. From the point after the introduction, the option to choose information about flight by time periods becomes available. Using a software package like Flowcharter ABC by Roycore, you can select sections of the linear presentation, cut, and

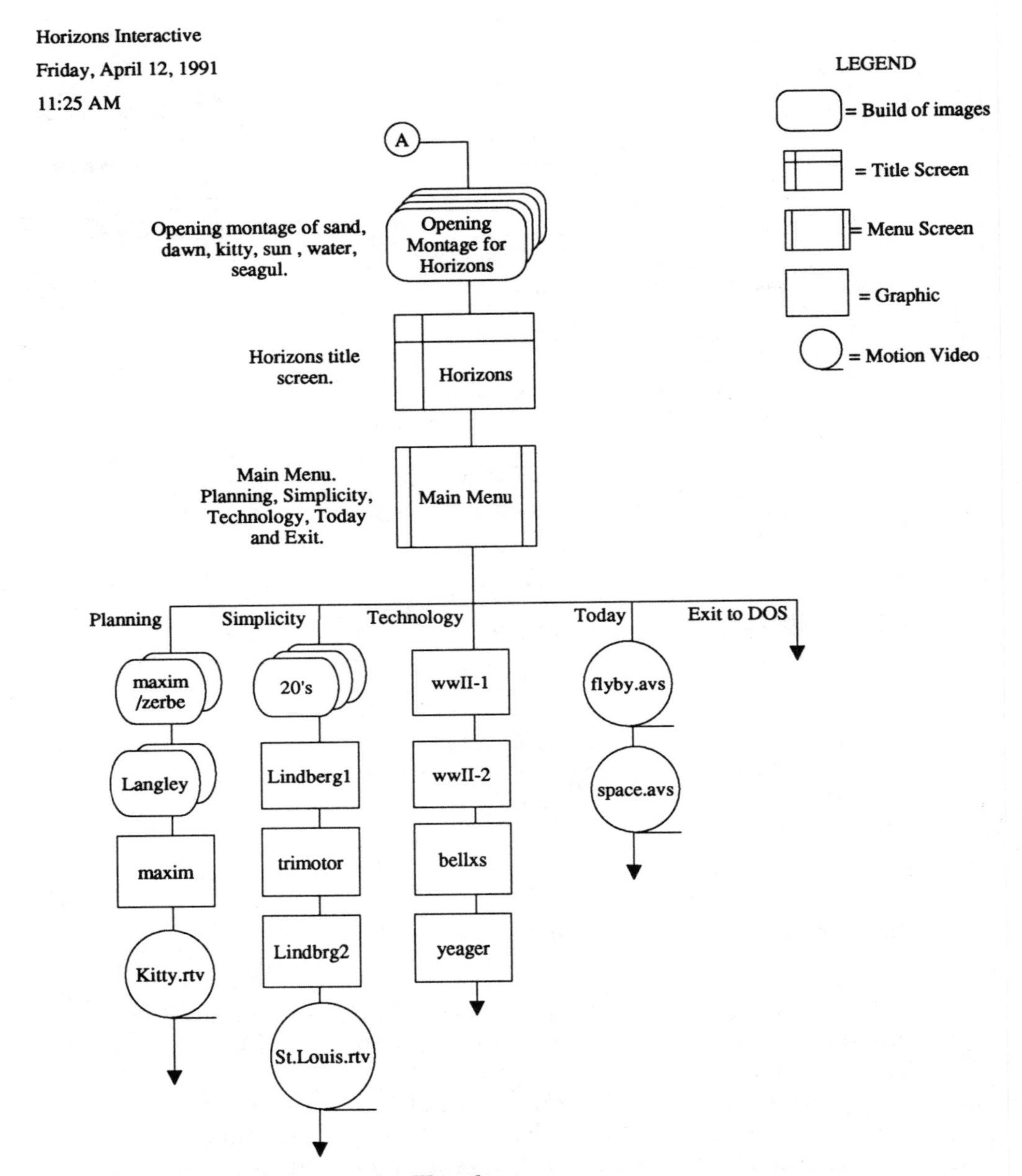

Figure 7.5 Simple Interactive Flowchart.

paste them into this alternative flowchart. This tool is easy to work with, and allows a preview of interactivity. Insertions can be made easily, as can deletions and changes—all the same reasons that we find using the personal computer as a tool for production planning so appealing.

These illustrations also help to show the need for *both* storyboards and flowcharts. Without the storyboards, the actual navigation devices—menu options, buttons, and the visual effect of how the information is presented—is lost. Flowcharts are the character; storyboards are the personality.

The flowcharts here show very simple applications that are either linear or have limited branching. One of the most powerful features of DVI® technology is the ability to give application users unlimited choices in the way that they approach or seek information. Some people have coined the word "hypermedia" to describe this feature. Essentially, this capability gives you the power to "link" certain information, whether it is a word in the text to a motion video sequence, or a still image to a series of stills that shows more detail about a particular topic. The encyclopedia example used earlier is one good example of this. While reading an article about semiconductor manufacturing, you may want to see a fabrication plant in action. By clicking on an icon, or a phrase highlighted in the text, a "video window" may appear on the screen and play back a video tour of the plant.

A flowchart for this type of an application will be very different from our two examples. Like the storyboard for this type of application, the flowchart will only identify types of interactions. The type of links would be identified, and a database of the elements and their respective links to other elements would then be attached to each of the flowcharts used.

The first example of a more complex application is shown in Fig. 7.6. This application is the interactive training software designed to help new LAN users learn more about the network. This application has a nested menu—the introduction leads you to the main menu. The dark shadow under the menu icon indicates that there is another, attached flowchart for that topic. One useful way to think about a software program is as if it has dimensions. The shadow under the menu icon indicates that there is more information "under" this icon.

This flowchart shows that there are four menu options. The choice of "Application Software" takes you into a linear presentation about software, and then to a quiz. Performance on the quiz may lead to remediation. The "LAN Operation" choice takes you to a short introduction and a second choice for another menu. This flowchart illus-

LAN2.ABC\TopChart

Training Module - App 2

Wednesday, July 18, 1990

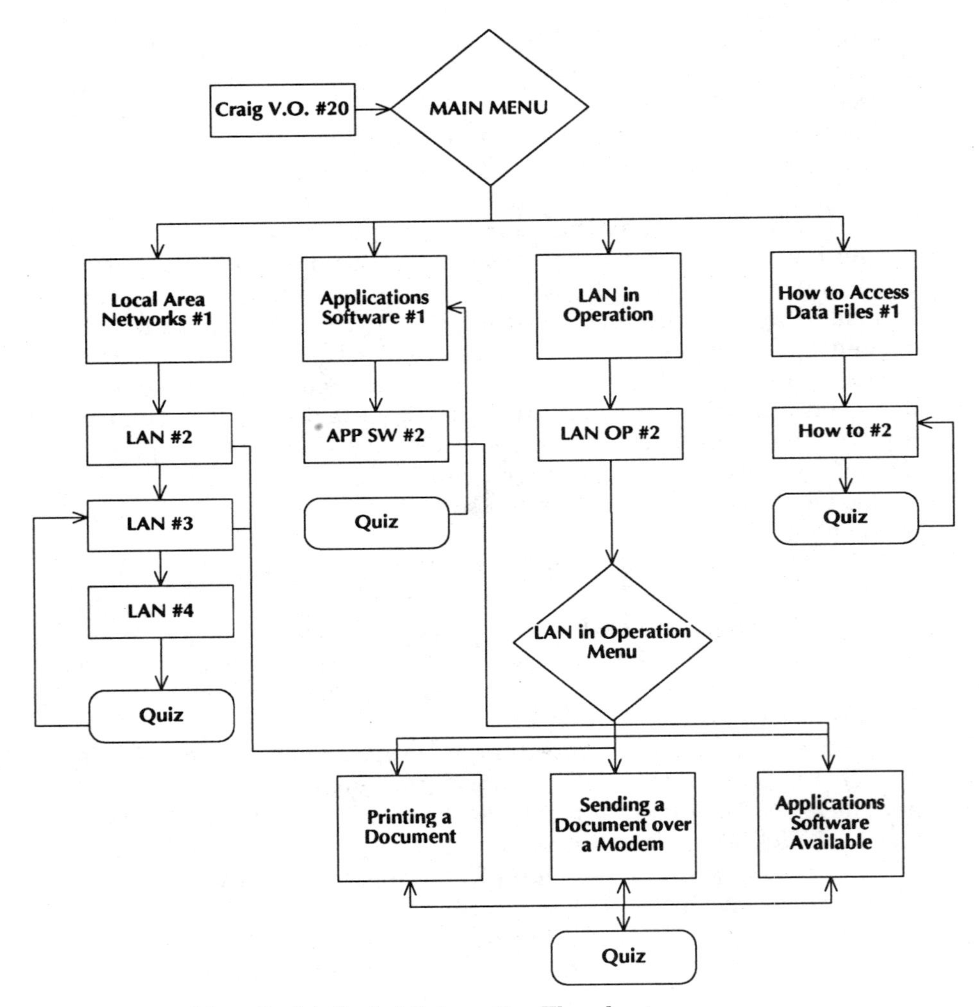

Figure 7.6 More Sophisticated Interactive Flowchart.

Figure 1.3 CableShare Real Estate Information.

Figure 1.4 Factory Inspection Information.

Figure 1.5 Factory Inspection Information.

Figure 1.6a Lindsay Library Town History Archive.

Figure 1.6b VideoFAX Medical Application Using DVI.

Figure 9.3 Horizons Opening Screen.

Figure 9.11 VGA and DVI Display Combined.

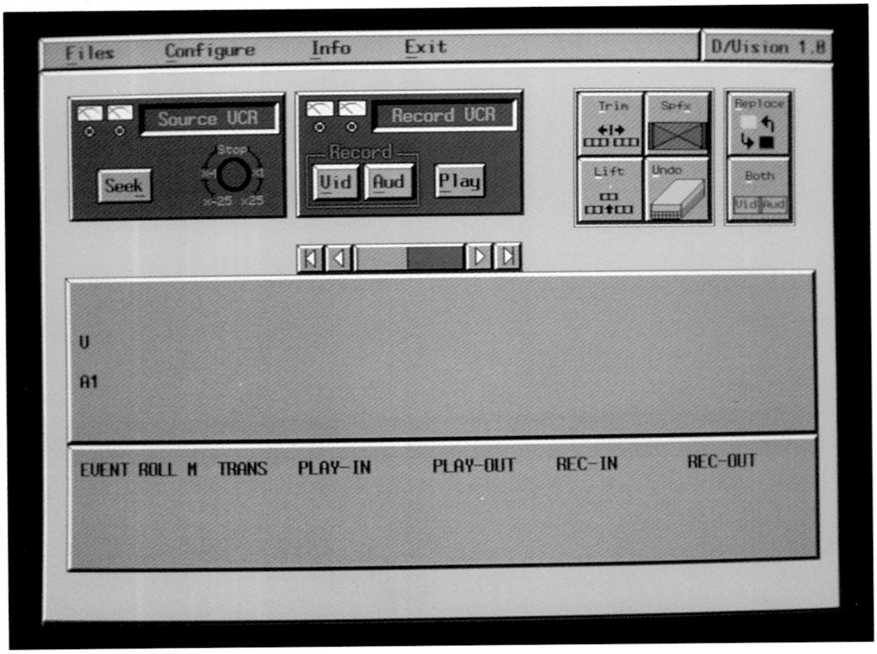

Figure 11.4b D/Vision Screen.

duction and a second choice for another menu. This flowchart illustrates a remediation path, and the use of nested menus.

At any point in this application, a user may come upon a term that will need definition. When a term is selected, the application will immediately show the term and its text definition. More information about any of these terms is also available. The information can be audio, graphics, motion video, or a combination of this media—usually taken from another part of the application itself. If the user selects a "More Information" option, the application program will automatically show the correct sequence. At the end of the sequence, the user is returned to a previous screen. All of this is documented on the flowchart, which the programmer used to make sure that these interactions were possible.

Our last example of an application that uses hypermedia concepts is one that was used as an educational piece to teach people about semiconductors. Figure 7.7 illustrates how you can use a topic, shown here as a graphic representation of an illustration, as the central menu choice. The "Chip Menu," or element "CN22," is a jumping-off point for more detailed information about the i486 microprocessor.

The flowchart identifies how a person using the application would get to this particular topic choice—in the upper left corner there is an arrow that indicates the user came from a screen identified as "CN2." The developers also provided information about how interactions would take place, and what essential information would be needed on a screen. For example, the flowchart designates that a "Go Back" button is needed on elements CN22–CN27. This illustrates how a storyboard and flowchart are so closely related. While this flowchart does not lay out specifically what that button should look like, or where it should go on the screen, it does identify the need to place a button on these elements.

Like many of the choices you will make in production, the type of flowchart you choose to use is one of personal preference. The two major formats we have shown—the branching, nested-menu format and the hypermedia example—could probably be interchanged. The difference is really in how these designers were thinking about their particular application. The LAN application had a very specific training goal, and needed to document success through a performance quiz and remediation option. The i486 portion of the educational application had less specific goals. The intent of this application was to give users an easy and accessible way to explore information about the i486. The approach to documentation reflects these differences. However, this should not stop you from mixing and matching these approaches as needed. You and your

TopChart

Wednesday, April 10, 1991

1:58 PM

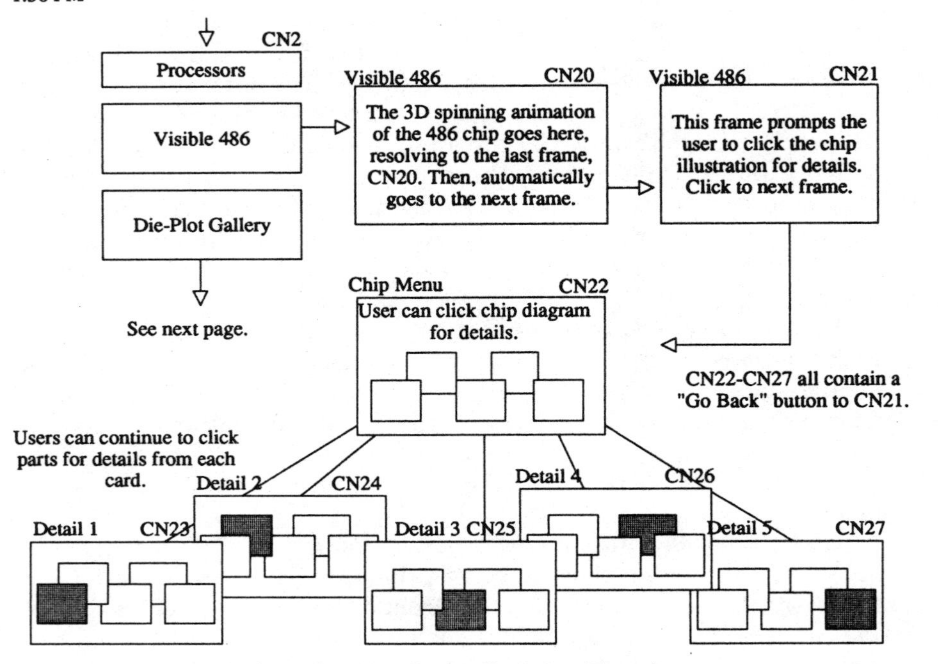

Figure 7.7 Example from Semiconductor Training Flowchart.

team should find the most comfortable, productive format for your applications.

Writing a script

A script for an interactive program is similar to a script that would be used in television or other motion video presentations. The script describes the music (mood, if not selection), the stills to be used, any special effects (if known at the time), and voice-over or actor lines.

Templates for script writing can vary, depending on styles of a particular writer. At the least, standard conventions are used to note the different elements. In most scripts, the audio and video are separated.

Video includes a description of any still images (image file names are used if available), and motion video. Video effects are also included—designated here in all capital letters. Audio includes the narration and any sound effects. A description of music, or actual names of pieces, would also be included here. The script may evolve, but at some point, the narration and actor lines have to be completed and locked in.

The script for a narrator or actor must be precise and complete, since production will typically happen only one time. Regular script reviews are needed, and if the client or management has to sign off on the script, a schedule that defines dates for delivery of first draft, reviews, and revisions are needed. The "shoot" or recording sessions are critical path items in the project plan.

It is also valuable to write "pick-up" lines as a part of the script for the narrator or actor. These lines or brief scenes do not have a particular defined use when they are scripted, but fit into the application's general tone and direction. Their purpose is generally to cover unanticipated branches, interactive choices, or to prompt the end user to make a choice. "Pick-up" lines are developed along with the storyboards and scripts.

Here are some example pickup lines:

For training applications:

"No, that selection is not correct. Let's review this subject in a different way."

"Very good. You answered all three correctly!"

"You have completed this session. Choose another section now."

"Make your selection now."

For point-of-sale applications:

"I am sorry we are out of stock on that product."

"Make your selection now."

"Make a selection now to see more."

It is best to make these lines generic, but specific to the functions of the application and the audience. In Palenque, a prototype application targeted for children, the pick-up lines were read by the young adolescent actor who played C.T., the central character. These lines had a slang-filled, easy-going tone. In a recent demonstration we completed for placement in Washington D.C., the pick-up lines are more low-key and characterless, used to prompt users rather than engage them with the personality of any character.

CONTENT AND CREATIVITY—MAINTAINING FLEXIBILITY

We like to think of all of the documents that are created as a part of the design process as "living" documents. They grow and change over the course of the product's development. It is important to define as many parameters as possible in production, without overdefining. Special effects, timings, still images, and graphics can evolve and change over the course of the project with little disruption. Even some content elements are likely to change as you put together an application.

This is also why we like preparing all of these documents on a personal computer. Updates can be quickly prepared, printed, and distributed to your production team and client or manager. An important part of each document is the revision number, date, and time prepared noted on each page. And an important part of revising this information is a review by each member of the team.

The elements within a DVI® application that can be changed during production for the least cost include:

Still images (if a variety of stills are readily available)

Text

Graphics

Graphic design

Timings

Music selection from music libraries

Special effects created in software

Changes that will cause more costly budget increases include:

Original motion video production

Motion video compressed off-line

Original audio production

Of course, any change, deletion, or addition must be weighed in terms of the final product. Choices and flexibility within the DVI® system are great advantages in design and development, but can also lead to a phenomenon known as "creeping elegance." Each addition or change should be evaluated against the original goals of the project, and against the budget. If the actor who read the script just does not meet the expectations of the team, for example, and it is determined that the product will suffer as a result, then the investment to reproduce it is wise.

In the next section of this book, we will provide you with a range of production tools, tips, and approaches that will help you make the best decisions about producing each media element for your applications. If your team can go into this next phase, the production phase, with a solid treatment, budget, time line, and the blueprints for content and design—flowcharts, storyboards, and scripts—you should feel well-equipped to divide and conquer the production tasks.

Section

2

Production

Chapter

8

How DVI® Technology Works

The theory of digital motion video and motion video compression has been introduced and discussed in a variety of places in Section 1 to familiarize you with DVI® technology and its hardware and software components. We are beginning Section 2 of this book with a more in-depth treatment of how digital motion video works in relationship to production. The i750 video processor offers multiple resolution and pixel depths for use in the design and production of applications, and the multimedia producer must be well-informed about how the options work together, and how they affect the look of the application.

In this chapter, we will review the basics of how an image is displayed on the PC, and provide details about image formats and resolutions, setting the stage for more detailed information and specific production techniques for merging motion video, stills imagery, and traditional text and graphics, all on a PC display.

DISPLAYING STILL IMAGES

Figure 8.1 is a block diagram describing the functional depth displaying an image. When a software program calls an image to be displayed, the file is transferred from a hard disk or CD-ROM to the memory of the ActionMedia display board.

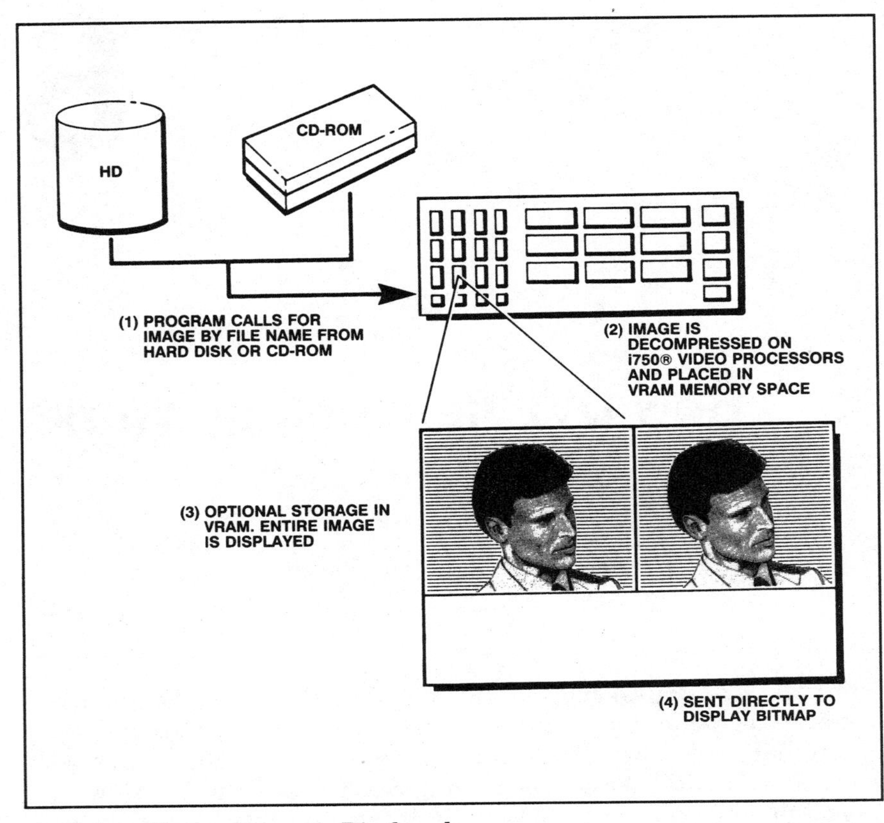

Figure 8.1 How an Image is Displayed.

Resolution

For example, a 512 × 480 still image consists of 512 pixels across (the X axis) by 480 pixels high (the Y axis). In the example, each pixel contains the information to turn a pixel off or on. The pattern of off-and-on pixels makes up graphical shapes and pictures.

Because the i750 video processor is programmable, there is more than one display resolution available for video imagery. Motion video, for example, is displayed in a resolution of 256 pixels across by 240 pixels high. Still images can be displayed at a variety of resolutions up to 768 pixels across to 480 pixels high.

Pixel depth

In our example, a screen image is made up of pixels in either an off or on mode. This is useful to introduce the idea of how images are

Figure 8.2 Image Displayed 2-bits Deep.

displayed, but is only applicable in a system with a monochrome monitor. Each character in a text screen is displayed as a matrix of pixels and graphics, but offers no shading or color to the image. When shading and color are added to a PC display system, we say that the image has *depth.*

The computer data used to describe the on or off state of a pixel requires one binary bit of data. In the lowest level of a computer's understanding, a bit is either represented as an electrical pulse being off or on or the mathematical equivalent of a 0 or a 1. An image where the color data is represented by only 1 pixel would be called 1 pixel deep.

Gray scale or multiple-shade images

Now, suppose we had the luxury of two binary digits to represent the status or color of each pixel. Two binary digits can represent a total of four different states. Literally, this would be represented as 00, 01, 10, or 11. This now means that our pixel could be either off or on, or could be one of two different intensities of light colors or shades of gray. This would allow an image to be made of a series of colors as shown in Fig. 8.2. This is still not adequate to display the detail of a black-and-white photograph, but provides more definition than the former example.

Four-bit color

The next step of quality assigns each pixel four bits of data, and now sixteen combinations of color and intensities can be displayed.

Figure 8.3 Image Displayed 4-bits Deep.

Sixteen shades of color, or black and white, are enough to define the image further (see Fig. 8.3). In fact, an EGA color image is displayed from a fixed set, or palette, of sixteen colors. In either color or black and white, this image is four pixels deep.

Eight-bit color

The next step in image quality is eight bits deep, and can represent 256 different shades of gray or color states. When a computer displays color in the eight-bit mode, the software must either set up a prespecified, custom color table of 256 colors, or use a color standard such as VGA. A custom palette is usually created by saving an image, and then mathematically selecting color values that exist in the image and are closest to one of the 256 colors available.

Sixteen-bit color

When the pixel depth is represented in sixteen bits, 32,000 colors are the result. Color is displayed by varying the values of separate red, green, and blue signals. Five bits of data are available for each red, green, and blue (RGB) signal, and one bit of color is reserved or overlay of text or other graphics on an image. Even the colors black and white can be displayed within the RGB mixture of colors. White is displayed when all RGB signals are at full intensity, and black when there is no signal at all, or a value of 0. The 16-bit mode represents a fairly complete color image.

Twenty-four-bit color

Many high-end color graphic display boards go one step further by using eight bits to describe each of the RGB colors. This allows for over 16 million color combinations for each pixel. In most cases, this is more color than the human eye can discern. For on-screen displays, it is more color than the phosphors in the monitor picture can display. For this reason, most multimedia image display modes stop at 16 bits deep. Desktop publishing applications utilize 24-bit color due to the wider range of color available in the print production process.

Pixels × color depth = file size

File size is a critical element when designing multimedia applications for both stills and to understand the particular challenge of displaying motion video from data played back from a CD-ROM. Now that you understand how the data to represent a picture is made up, calculating the file size is straightforward math.

512 × 480 × 1 bit deep = 245,760 bits ÷ 8 = 30,720 bytes or 311K bytes

512 × 480 × 16 bits deep = 3,932,160 bits ÷ 8 = 49,520 bytes or 492K bytes

512 × 480 × 24 bits deep = 5,898,240 bits ÷ 8 = 737,280 bytes or 737K bytes

DISPLAYING DIGITAL MOTION VIDEO IN A MULTIMEDIA SYSTEM

Displaying motion video on the PC offers a new challenge. NTSC, a standard for video display, displays motion by showing 30 unique frames of video every second. Using the simple mathematics shown above, 30 frames of video at a resolution of 512 × 480 would require over 22 megabytes of storage for every second of video, and 1.3 gigabytes per minute. This is not a practical file size to store on PCs, and is beyond the data-transfer rate of a personal computer or local area network.

Research conducted in the transmission of color images for color television (see box next page) shows that the human eye is more sensitive to detail in a picture than to color. Further, image compression techniques for the PC can take advantage of the techniques used in transmitting color television signals. For color television, two planes represent each image. This image format is called YUV, where the Y plane is the monochrome (black and white) or luminance component, and the U and V plane represents the color values for a frame of video.

For further reduction of image size, using a high-speed video pro-

THE DEVELOPMENT OF DVI® TECHNOLOGY—A BIT OF HISTORY

The David Sarnoff Research Center was the birthplace of color television, and was the only research and development facility for RCA Corporation. The combination of scientists who understood the principles of color television, along with those who could harness the power of microprocessors, invented what is now known as DVI® technology.

The challenge taken on by a research team at the David Sarnoff Research Center in Princeton, New Jersey, in the late 70s was twofold. This team's goal was to integrate full-motion video into an all-digital multimedia system, the personal computer. First, motion video file sizes had to be reduced so that they would be practical to store. Second, a storage device that would be used in the next decade (this work began in the 70s) would have to be selected.

David Sarnoff Research Center researchers had already determined that the human eye was more sensitive to the detail of a picture represented in black and white than to color information. This work had been completed in conjunction with research on color television. It was this research that guided the work in image compression and decompression. Arch Luther's book, *Digital Video in a PC Environment,* is an excellent resource for more about the history of DVI® technology.

cessor, the color plane of an image can be sorted and averaged where only the color values for every fourth pixel are stored. This process is illustrated in Fig. 8.4. The result is color, full-motion images represented in an average of nine bits per pixel.

Each of the YUV planes is represented as eight bits per pixel (see Fig. 8.5). This provides all of the detail of an image. For the color data, three out of every four pixels are discarded in the two separate U and V color planes on both the X and Y axes to reduce the file size for each color plane to $1/16$ of its size.

Mathematically, this concept is illustrated as follows:

The image starts as:

$$256 \times 240 \times 16 \text{ bits deep} = 983{,}040 \div 8 \text{ bits} = 123\text{K bytes}$$

This is reduced to:

Y Plane – $256 \times 240 \times 8 = 491{,}520 \div 8 = 61$K bytes
U Plane is $1/16$ of data = 4K bytes
V Plane is $1/16$ of data = 4K bytes
$256 \times 240 \times$ average 9 bits = 69K bytes per frame

Using CD-ROM as a delivery medium

Since CD-ROM is a logical target for the distribution of software using multimedia (see Chap. 2), the goal of any compression of motion video must take into account the 150 kilobyte-per-second data rate

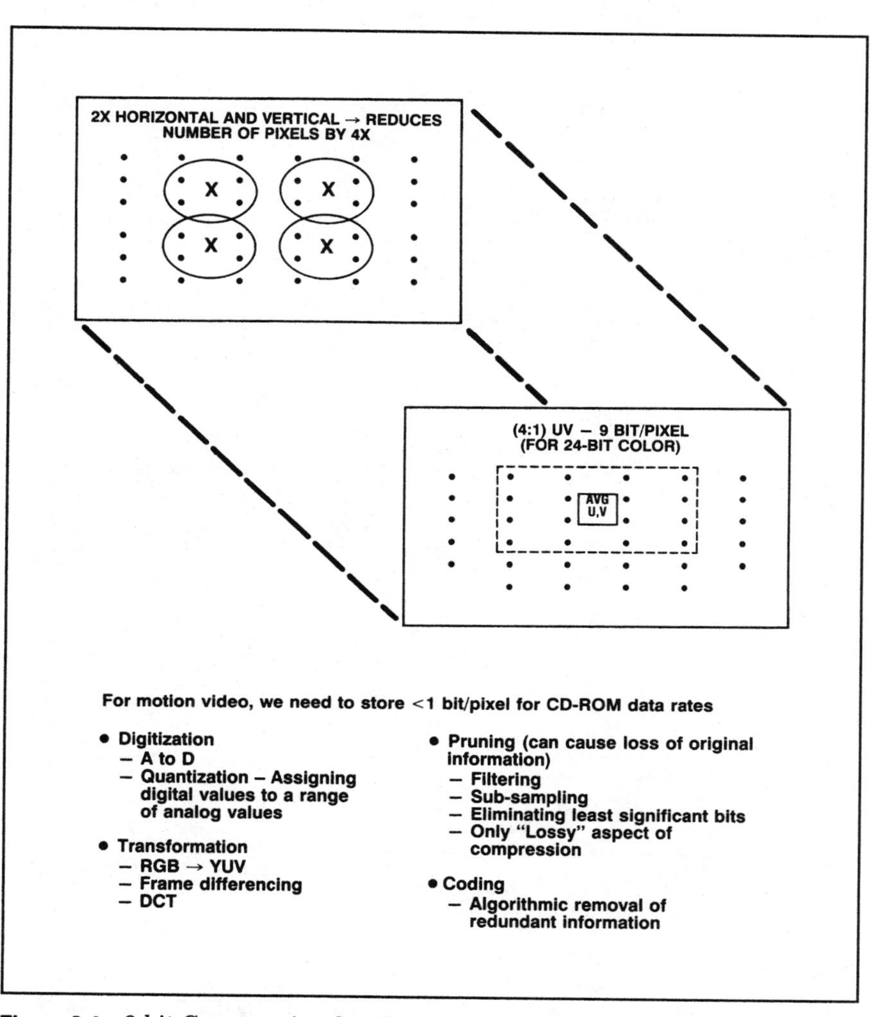

Figure 8.4 9-bit Compression for PLV.

from CD-ROM. This averages 5K bytes per frame. Other compression techniques can be applied to motion video to help achieve this goal, including a method called *run-length encoding.*

In a typical frame of video, there may be a continuous row of pixels that is the same color that represents the blue in the sky. It is more economical to represent this data by defining the color of the first pixel and copying it X pixels across, rather than repeating the color definition for each pixel. This method, along with delta encoding (described below), further reduces the size of the motion video frame.

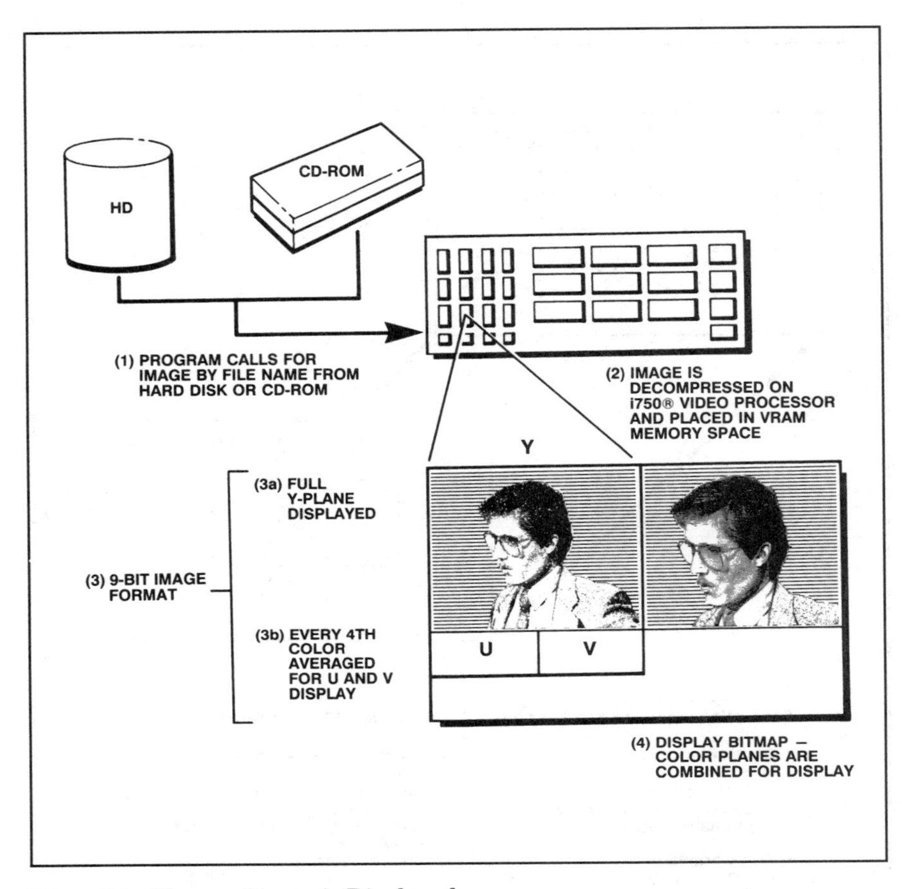

Figure 8.5 How an Image is Displayed.

Delta compression

If you recall making flip animation books, then you remember that to get the effect of motion, you only change one small part of the picture to make it look as if movement is happening. If you analyze a motion video clip, you will find that the number of pixels that change from frame to frame is also quite small. For example, often only the foreground action is changing with the background remaining static. By updating only those pixels that change from frame to frame, the data storage required to represent a clip of video is dramatically reduced.

As discussed in Chap. 2, a PLV motion video clip begins with a full 256 × 240 × 9-bit run length encoded image. Subsequent frames contain the image data that has changed from the previous frame. Full, compressed frames are periodically inserted when almost all of the pixels change (at the beginning of a scene change, for example).

The task of decompressing the data in a PLV file, and averaging the color based on a series of mathematical algorithms, falls to the i750 video processor. The i750 video processor was specifically designed for controlling the decompression and display of successive frames of video. The individual frames are decompressed, and the display is stored in VRAM in a series of frame buffers. The frames are then displayed in sequence, just as they are in any motion video technology.

Today, the off-line PLV compression process combines all the methods discussed here, and applies the power of a parallel processor to optimize compression and quality of motion video. An average data transfer rate of 5K bytes per frame is maintained for playback from a CD-ROM. 4500 bytes of each frame is for video information. Five hundred bytes are reserved for the audio data stream.

The technology of image compression, data transfer in the personal computer solved, and the i750 video processors are just part of this story, however. Boards that integrate the components were manufactured and tested, and more importantly, a set of system software, software libraries, and production tools are needed to bring this technological solution into a set of useful products for software developers and PC users.

While these tools, and the techniques we will share with you relating to the production of multimedia, are usually very specific to the individual media elements, it is important to note that all of the compressed media elements are interacting in a total hardware and software system. Some production techniques that become possible in this environment are unique because they combine elements; others are unique because of the complete integration of motion video in an all-digital system. In subsequent chapters, we will cover a number of them including topics like VGA graphic display, and the flexibility of different video display modes.

IMAGE COMPRESSION AND THE MULTIMEDIA PRODUCER

The programmability of DVI® technology presents the multimedia producer and designer with enormous flexibility. Small file sizes reduce the amount of storage so that multimedia can be distributed cost-effectively, and so that the speed and performance of the application is acceptable. When an application is distributed on CD-ROM, or over a network, the issues of file size must be considered as a high priority in the design process.

In addition, DVI® image formats and resolutions will affect how you produce the elements you eventually will use in your applications. For example, the PLV compression process assumes that in most cases every pixel is not changing from frame to frame. PLV and RTV video files are always displayed in 9-bit mode, at a resolution of 256 × 240. This has some implications for production of full motion video. Tips and guidance for shooting motion video for the best results for digital compressed video are covered in Chap. 11.

Likewise, still image compression also has implications for design and development of a multimedia application. Still image modes for DVI® technology include 9-bit and 16-bit mode, and a variety of resolutions. These issues are presented in more detail in Chap. 10.

Chapter

9

Designing Your Applications: Graphic Design for Multimedia Applications

The graphic look is a key element to the personality of an application. In the eyes of the viewer or end user, it sets the mood and creates subtle impressions throughout the application. The graphic look can dramatically alter the direction of the content within an application. A business information or training program will have a very different look from an entertainment game or a point-of-sale application for women's clothing.

Beside the subtleties built into the graphic design, it can also determine how easy or difficult any application is to use. An integrated part of the design is the navigational interface built into an interactive application. The design may incorporate buttons, menus, hot-touch spots, slider bars for viewing text, or icons to represent the various additional information resources an application may include, such as a clip of audio to support the on-screen visual.

The graphic elements that make up the look of an application include the backgrounds, textures, colors, and the way type is displayed on screen through the selection of fonts, style, color, and drop shadows. These graphic elements make up the design of an application and develop impressions within the mind of the end user.

The graphic design of an application is also carried through to how stills and video are presented. The complementary colors of the design may be used in stills photographed for an application. Or, the lighting of a sequence of video may be designed to match the look and feel of the rest of the production.

The job of bringing this all together and creating the design out of thin air falls to the graphic designer or creative director. The graphic designer or creative director has the eye and the knowledge to understand how to influence the end user's mood and feeling through the visual senses.

Where does the process start?

The process of developing a graphic design for an application starts during the concept stage of production. The graphic designer or creative director is typically included in early meetings when a production is first being discussed. The process for developing the design starts with understanding the application and its intended market. The subject matter for an application and existing graphic or visual elements will all begin to mix together within the eyes of the graphic designer to develop the rules for an overall design. The best way to understand this is to take a look at a practical example.

For an application called "Horizons," shown in Figs. 9.1, 9.2, and 9.3 (see color plate), the graphic designer looked at the subject matter

Figure 9.1 Horizons Graphic Look Example.

Figure 9.2 Horizons Graphic Look Example.

and began to develop the design for this multimedia module. The theme included a historical perspective on aviation. To add to the historical feel, old photos were used. These photos actually became the basis for the graphic design. To add the feeling of age, and a "looking back in time" atmosphere, the backgrounds and photographs were given a sepia tone. Borders were added around the pictures and a one-quarter screen window was used to present historical film clips. Working from the old-time etchings of early flying machines to photographs of the Wright Brothers and Lindbergh, warm brown and sepia tones became the dominant colors. The graphic designer then moved to the introduction of the application. To catch the eye, the warm sepia tones were converted into oranges and reds to provide a bold, compelling look for the title graphics.

The warm sepia colors are carried throughout the periods of time where black-and-white photos are used. In the application, when the time line begins to describe events of the late 40s and 1950s, when color photography was available, the colors and textures used in the backgrounds and type begin to take a warmer hue, taking on the pastel tones typical of photography of that era. This is a good example of how the graphic look and design complemented the visual material and created the mood intended by the creative team.

ELEMENTS OF THE DESIGN PROCESS

Let's look at the design process, and see how it might affect each of the different media elements available using DVI® technology. The

Figure 9.4 16-bit Marble Example.

ability to present still graphic images is one of the key elements of multimedia presented on desktop personal computers.

If we started with a blank screen we would want to add a background. This background could be as simple as a solid black screen or another, more interesting solid color. To add depth, the solid color can be graduated from dark to light. The designer can also take the next step by introducing a bit-mapped texture into the background such as marble. This bit map is an image file, and can easily be created by using a scanner or video camera connected to the ActionMedia capture board.

The number of things that can be digitized to create interesting backgrounds is surprising. Crumpled up paper or tin foil dramatically lit can be captured under a camera. Using Lumena for DVI® technology, color can be added or the contrast enhanced. Edges can be sharpened and softened and mundane objects, such as a paper towel or square of ceiling tile, can become an interesting texture for an application.

Marble has become very popular as a background texture for on-screen presentations. The ability to display images using 16-bit color allows the richness of marble to be displayed on screen. An example of this is shown in Fig. 9.4. Marble samples can be obtained from any marble wholesaler or supplier and digitized using a video camera or scanner. You may find it necessary to alter or adjust the digitized image of marble, as often the detail will be too sharp and will need to be softened so that foreground graphics are not overpowered.

Creating backgrounds using graphic elements

Backgrounds can also be created using geometric shapes, color, and even text. For example, a client's logo can be digitized, shrunk, and duplicated across the screen in a grid effect to create a wallpaper-like background that will not conflict with foreground graphics. Or, a graphic element from the logo can be used to tie the graphics together as shown in Figs. 9.5 and 9.6.

Geometric shapes such as triangles, squares, and circles can be colored and shaded to create background effects. Using deep shadows can create a feeling of depth.

Backgrounds from other computer platforms can also be used. A design can be created using the unique graphics capabilities of one of the many computer graphics software packages available and converted using the VImCvt software utility available in the ActionMedia Production Tools software.

Ruled lines and different textures as part of the background design

A background design does not have to be a wallpaper pattern displayed edge-to-edge. A ruled line across the top one-third of the screen may denote an area for a headline copy to the screen. The line can be displayed in a complementary color to the background and

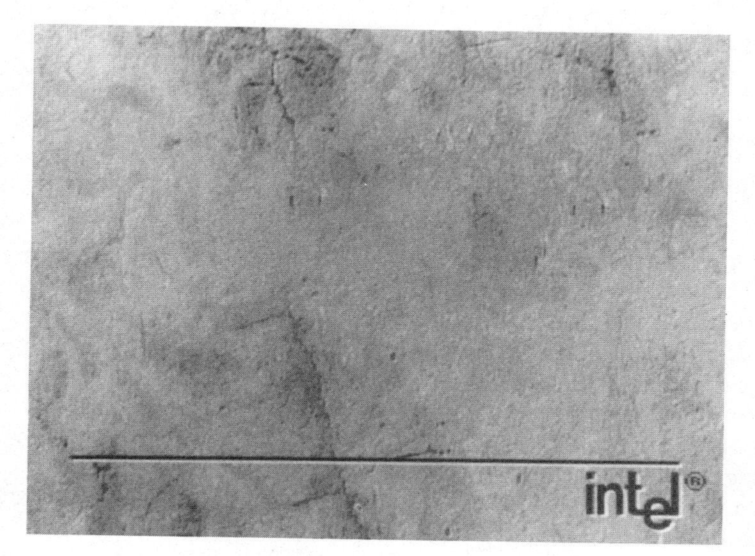

Figure 9.5 Logo with Marble Background.

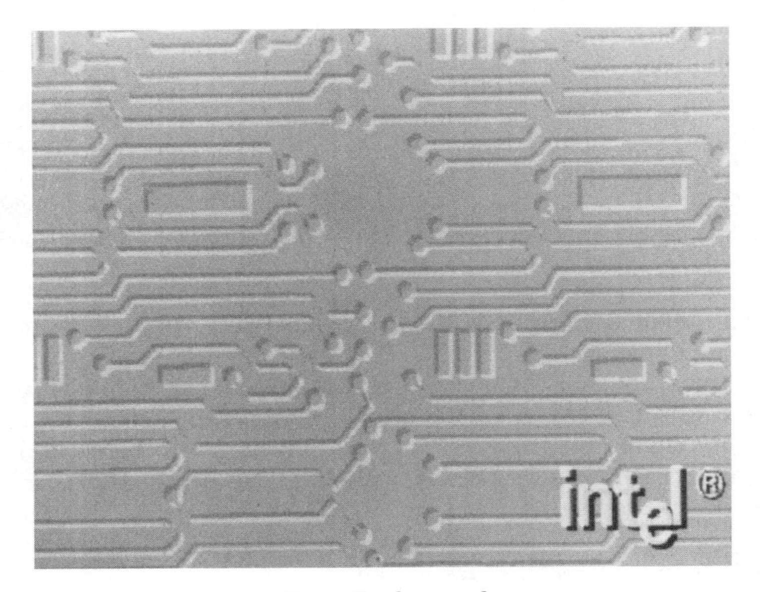

Figure 9.6 Logo with Circuit Trace Background.

may include a drop shadow behind it or a light or dark-colored beveled edge to add depth. A series of ruled lines at various lengths is another idea to further complement the design and delineate different sections or the bottom margin.

Different textured areas of the screen may be defined. The "Creative Ideas in Action" screen shown in Fig. 9.7 is an example. Here, the marble bar is used to stand out from the background as the area for the presentation's logo and navigational buttons.

Photographs as backgrounds

Photographs and still images can also be great graphics for background for an application or part of an application. Using Lumena for DVI® technology, a photograph can be converted, contrast added, or it can be posterized to create an artistic effect for use as a background on an application. By experimenting with images and various capabilities of a paint package such as Lumena for DVI® technology can create many interesting effects.

The key to developing a background is to keep in mind that it is designed to complement and not overpower the images, text, and graphics in the foreground, and should complement even the narration and audio effects.

TEXT AS PART OF THE GRAPHIC DESIGN

As with the design of the background, the selection of fonts starts with understanding the intended application of the program. Fonts, like the background effect, can enhance the mood or image of the presentation.

To use an exaggerated example, a multimedia presentation for Rolls Royce Motor Cars would probably utilize a serif typeface with its classic lines, whereas the Jet Engine Division of Rolls Royce might be inclined to use a sans serif modern face such as Helvetica for its technical presentations. When a designer begins to make the decisions that create the image of an application, the font selection begins with a choice between serif and sans serif typefaces. The next decision based on the typeface typically requires a review of the fonts available. A family of over forty fonts is available for creating still images with Lumena for DVI® technology, and for display on-screen using font outlines from Time Arts and the font engine utility included with Intel's ActionMedia 750 Production Tools. An application will probably have several font needs. The three main uses of text are:

Headline or display font

Text font

Font for emphasis

Figure 9.7 Creative Ideas in Action Logo.

Also, various sections may have slight differences in their presentation requiring a change in the graphic look which can be designed into the application through the effective use of different complementary font styles.

Headline or display fonts

Often a headline or display font is in boldface, and is used to summarize the other text on the screen. A headline or display font can sometimes be very different from the text font. For example, a combination of Helvetica Condensed, a sans serif font, can work very well with Times Roman as a text font throughout the application.

Text fonts

Selection of the text font carries its own rules. Many of the rules are common sense and come out of the principles of good graphic arts design. Because by nature of the physical space on the screen, the graphic designer has to make a decision on the size, spacing, and the maximum number of lines of text per screen. If the application must be designed for a lot of text to be presented on each screen, a sans serif face may be better. The sans serif face will look better when presented in a smaller size, and allows more lines of text on the screen.

Using italics for emphasis

The italics style of font is often used for emphasis in print design. Depending on the size of the typeface presented on screen, this can work for multimedia presentations. Be aware that italics fonts are more difficult to render in on-screen presentations because of the slanted presentation. Because fonts are rendered as a matrix of horizontal and vertical pixels or dots on the screen, an italics font may show the jagged edges of an angled line more than normal or boldface styles with their primarily vertical and horizontal orientation.

Other factors when selecting fonts

Color is another consideration. The color of the fonts must work to complement, but yet stand out from, the background. Certain colors are easier to read on screens, with the best being those that offer the most contrast, such as black or white. A gray font, though it may look the best graphically, may be difficult to read. The graphic designer has to carefully consider the use of color and weigh it against the user's ability to easily read it.

FONTS DYNAMICALLY DISPLAYED USING DVI®

DVI® includes a method for creating system fonts that can be used for dynamic display within an application. Both the ActionMedia Software Library and authoring products such as Authology: MultiMedia and MEDIAscript can be used. By using the font engine software, bit maps of display fonts can be generated for an application. Using this technique, the font engine is used to create a file of bit maps which includes each character of the alphabet. The font engine requires several parameters to create a font set such as the font face, style, size, and pixel depth.

Once a designer has selected the fonts for an application, the necessary bit map files are created using the font engine. When an application is loaded for playback, the font bit map files are preloaded into the video memory of the ActionMedia Display board. As the application is played back, the bit map of the individual characters is displayed on a character-by-character basis on-screen to make up lines and sentences of text. This is the preferable method to developing an application, as it will save a significant amount of data storage space.

Using dynamic bit map fonts is preferable to the alternative method of storing each text frame as an individual bit map for the entire frame created using Lumena for DVI® technology. The graphic in Fig. 9.8 required 136K of data storage as a high-quality DVI® compressed, 16-bit file, or c16 file. If an application's design required 300

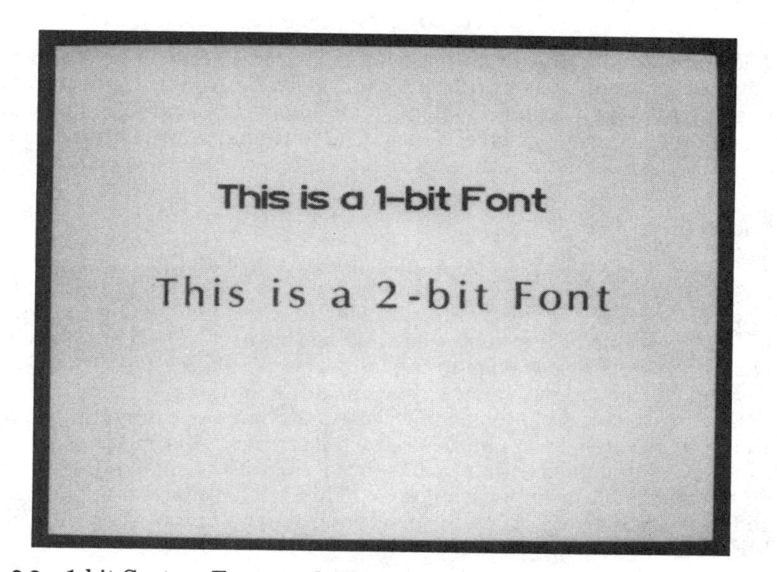

Figure 9.8 1-bit System Fonts and 2-bit Dynamic Fonts.

such text screens at an average of 256K each, the total storage required for the text screen would be over 40MB of data.

Using dynamic fonts, this can be reduced, as only the background needs to be displayed once. The individual text can be displayed on each screen from simple text files and graphic calls with almost the same effect. The difference would require storage of the background file (136 kilobytes) and font bit map files (three faces for a total of approximately 275 kilobytes). Each screen would also require about two kilobytes for the storage of the text data in string form for presentation on that screen. Using dynamic fonts, this example would require less than two megabytes of data storage.

This is a significant design factor to consider. An application requiring two megabytes of data could be distributed on floppy disks. A 40-megabyte file forces the need to distribute it on a high-density, low-cost media, such as CD-ROM.

For distribution over a LAN, this is also important. Over a LAN, the background file and font bit maps can be sent from a central file server once with each individual screen only requiring its associated text string data. This efficiency is smart planning during the application design. The technique of using bit map dynamic fonts also enhances the response time of the application. Once a background file and the font bit maps are loaded into the ActionMedia board's memo-

DVI® DYNAMIC FONTS

The ActionMedia 750 Production Tools contain a font engine and font outlines to produce Times Roman and Helvetica fonts. Using the font engine, the application developer can select as parameters the font typeface, style, size, or pixel depth. The font engine then creates a bit-map file from the font outlines based on the parameters selected.

Two-bit fonts

The term *two-bit font* is very similar to the concept of graphics presented two bits deep. Single-bit fonts only describe the outline in either full-color or no-color. By describing the font in a total of two bits for every pixel, two levels of luminance or shading are used. These two levels of shading give a smooth edge to the font. The results are particularly evident on characters with curves, such as an "s" or an "o."

One background can be used throughout your program, and the text can be generated from simple character strings. This means that text can be updated easily. The text is not a part of the image file, but can be changed in the application code. This is easily accomplished in C code, or with the authoring products, Authology:MultiMedia and MEDIAscript.

ry, the graphics processing and display time for successive screens of high-quality display text is minimal.

One-bit or two-bit dynamic fonts?

At this time, DVI® supports both one-bit or two-bit bitmapped fonts in the DOS environment. The terms one-bit- or two-bit-deep fonts describe the level of antialiasing available for each character. Antialiasing is the process of reducing the jagged edges around the curves of the character represented on-screen as a matrix of pixels. This is especially important for curved or rounded characters or serif fonts. The term one-bit-deep means that the pixels which make up each character can only be represented to the computer's binary thinking as either off or on, the equivalent of a one or zero. Two-bit fonts offer a total of four different states: off, on, and two middle tones of gray. The middle tones of gray are important to the presentation of the curves of the character, as the rounded edges can be represented with shading, giving the appearance of a smooth edge. The more shades of gray or pixel depth available, the smoother the rounded edge.

Graphic screens created using Lumena for DVI® technology and saved as a complete image can be represented on-screen with antialias fonts four to eight bits deep. This represents the best quality presentation of text and fonts. This difference in quality comes at the price of requiring significantly more data storage for those applications which will have a series of successive and separate still frames utilizing the same fonts. The choice is up to the designer, with the tradeoff being better performance and significantly less storage for an application using dynamic fonts.

STILL IMAGES AS A DESIGN ELEMENT

An application can be designed to display all still images full-screen. For many applications, this may be required to show the maximum amount of detail. Unfortunately, after a while, an application designed solely with full-screen images will appear more like a slide or film strip presentation. A graphic designer can add interest by varying the presentation of images in different sizes and positions on the screen and mixing them with the background.

The graphic designer can also add a framing element to an image. This may be as simple as a black ruled line to differentiate the image from the background. Or, a drop shadow can be added to make an image appear to float over the background.

A bevel-edged border can be added to the image to make the image appear to either float on a higher level above the background, or be recessed and framed into the background, depending on the placement and shading of the shadowed areas of the bevelled edge.

Using images as a special effect

An interesting graphic design technique is to pull the image through the background. Here, a part or all of an image is converted to a percentage value and placed over a background. For example, in Fig. 9.1, shown earlier, the image of Lindbergh is pulled through the background at a value of 60 percent of its original density. This allows the high-contrast areas of the image to show through the background with the lower density or clear areas carrying the texture of the background. To create this effect, the original image was modified to increase the contrast, and all of the original background was airbrushed out. Then it was combined with the background, and placed in the appropriate position over the background. The result is a subtle graphic design that can support narration or become part of a background that evolves throughout the audio presentation as it does in this multimedia application.

Using stills to create the impact of motion through transitions

Many film documentaries have used a series of stills in place of motion video to tell a story. This has often been used for historical subjects that took place before film or video was invented. ActionMedia software and the authoring products include a number

USING COLOR LOOK-UP TABLE GRAPHICS

In many instances, the design of an application will call for display of images in a partial screen. Figure 9.10, a sample screen of a point-of-sale application, is a perfect example of this. The technique used here is a marriage of a traditional image-display technique, *Color Look-Up Table* (CLUT), and DVI® still-image format. The background for this application was created in the 256 × 240 CLUT mode by creating a custom palette. In CLUT mode the palette is limited to 16 colors, with 16 additional levels of luminance for each color. Using a sophisticated paint package such as Lumena for DVI® technology, interesting graphic backgrounds can be created by using textures.

As we have noted earlier, DVI® offers a lot of flexibility. One of the trade-offs to using motion video is the need to be in 9-bit mode for easy transitions between stills and motion video. Unfortunately, the pixel averaging used to compress motion video to 9-bit mode can cause the colors of text to bleed against the background. The CLUT mode resolves this problem. With CLUT mode, text fonts can be displayed with no bleeding against a colored or textured background. As shown in Fig. 9.11 (see color plate), 9-bit images can be copied or overlaid on the background with clean transitions to motion video.

VGA DISPLAY

Another mode available to the DVI® multimedia producer/designer is the use of the VGA graphic display. If you have a VGA display in your computer along with ActionMedia hardware, VGA can be used in its normal mode, or a smaller window on the VGA display can show DVI® video (PLV or RTV). The designer also has the option of switching between full-screen DVI® display or the full-screen VGA display.

Figure 9.11 (see color plate) shows an excellent implementation of VGA, together with DVI® display, in an interesting version of interactive tic-tac-toe. Here, the game-board borders and scores are displayed in VGA mode. The VGA display is at a resolution of 640 × 480, and is mixed directly with the PLV motion video, the software program that controls the application. Note that the X's and O's are displayed graphically in VGA form. This interesting technique opens up many opportunities. For example, a DVI® window could be used in one corner of the screen for training or audio-video help information while the VGA application is displayed. Motion video or stills with audio could be used to assist an end user with a short tutorial.

of routines and transitions that can be combined to create a very interesting "motion presentation" using only stills.

At its simplest, a presentation could be quickly designed and developed to appear like a film strip presentation with digital audio complementing the full-screen stills. But, with a little extra effort and a knowledge of how to use the available tools, the screen can come alive with effects and transitions to carry the impact of a multimedia presentation.

For example, if you go back to our example in Fig. 9.1, a still image was pulled through the background to create a new image. You could take the image of the background without the image and the later one with the image pulled through and save them separately. Now under program control you have the choice of transitions between images. A "cut" could be accomplished by just displaying one picture right after the other. The effect would be the sudden appearance of the pulled-through image of Lindbergh. A more effective and esthetically pleasing design would use a dissolve, wipe-on or other special transition effect, such as a "venetian blind" wipe. In our example, because the two image backgrounds are the same, the transition effect makes the image of Charles Lindbergh an addition to the existing background.

Transitions are where the fun and creativity really begin for the graphic designer. With print design, the entire message must be communicated graphically within the bounds of the single dimension of a page of an advertisement or a brochure. With multimedia, the graphic designer has the capability to make a single piece of art come alive

COMBINING DISPLAY RESOLUTIONS AND SWITCHING DISPLAY MODES

Motion video in an application must be displayed using 9-bit mode, while stills can be displayed as 16-bit or 9-bit (see Chap. 10). If the design of an application calls for transitions between a 16-bit still image and motion video, the mode of display has to be changed through the software program. The visual is changed on-screen from 16-bit to a duplicate 9-bit image. This results in a softening of the image due to the pixel averaging techniques used in 9-bit mode described earlier in this chapter.

More importantly, our screen display is now in a 9-bit mode. As you can see in Fig. 9.10, we can now display 9-bit motion video. The example in Fig. 9.10 actually combines two design and production techniques. The motion video in this screen is being displayed in the higher-quality 512 × 480 resolution mode. Note that the motion video is displayed in one-quarter of the screen. Since the display resolution for motion video is 256 pixels across × 240 pixels high, and this is exactly one-quarter of a 512 × 480 screen, the image is shown as a quarter frame.

The result of this technique is a very smooth transition where still-image artwork, menus, and backgrounds can be maintained and combined into the display with a "window" of motion video. At the end of playing the motion video clip, the same technique can be reversed to return back to the original 16-bit image. The added benefit to displaying motion video on one-quarter of a 512 × 480 screen is that the quality will appear higher, since the pixels for the motion-video file are only spread across one-quarter screen area.

into motion with transitions. It starts with the background, and can continue to flow and build from a background which sets the style and tone to the addition of images to support the narrated story with pictures.

Adding audio to a graphic design

In Fig. 9.9a, we see the beginning of the Horizons title screen sequence. Unfortunately, in a book one cannot hear the soundtrack which starts with the striking chords of music and the sound of seagulls mixed with the sounds of the surf. The seagulls are built on the screen in several venetian blind transitions to build to a sky full of birds (see Fig. 9.9b). As the sound of a small gasoline engine sputters on the soundtrack, we begin to see the image of the original Wright Flyer plane appear on the left side of the frame (see Fig. 9.9c). The music begins to swell as the sound of the Wright Flyer trails off and is followed by the rush of a jet taking off. Here the shadow of the jet wipes on center screen followed by the type treatment for the title of the show—Horizons (see Figs. 9.9d and 9.9e).

Figure 9.9a Horizons Title Screen Sequence.

Figure 9.9b Horizons Title Screen Sequence.

Figure 9.9c Horizons Title Screen Sequence.

Figure 9.9d Horizons Title Screen Sequence.

Figure 9.9e Horizons Title Screen Sequence.

Figure 9.10 One-quarter Screen Motion Video in 512×480 Display.

This is just a short example of how the design of the visuals, including the background, individual graphics, and photographic elements, were brought together into a series of images combined with a soundtrack. During this section, the design did not include extensive motion video. You can see how the effective designer, using transitions to build images on the screen, can create the same feeling of motion in the eyes of the user.

The key is in the design of the application and the knowledge of what multimedia presented through DVI® is capable of doing. The next several chapters will provide the building blocks to learning the capabilities. With this background, the graphic designer can design multimedia applications that will keep the viewer visually involved with the subject matter being presented.

Chapter

10

Capturing Still Images

USING STILL IMAGES IN AN APPLICATION

Still images can be a powerful element for multimedia, especially combined with the audio and special effects available using DVI® technology. Still images are also important to the overall design and development process of application development. They can be used in the creation of storyboards, and can help to communicate effectively with a client or manager.

Still images are among the elements that can be changed in the application with minimal effort and expense if the content is available. If the design team or client finds a still image that is an effective addition, it can be added to the application without expensive production. If one is not effective, it can be replaced easily. Still images in a DVI® production environment are digitized, compressed, edited, and displayed all on the personal computer equipped with ActionMedia hardware and software.

The flexibility of the DVI® development environment introduces many options for the capture and display of still images. We have devoted an entire chapter of this book to still images for two reasons: Stills can be a valuable resource for communicating, and the ActionMedia 750 Production Tools allow many options for stills that can have a direct impact on the application's quality and budget.

REVIEW OF THE ACTIONMEDIA 750 PRODUCTION TOOLS

ActionMedia 750 Production Tools are the software component for capturing, compressing, and editing audio and video elements. Production tools are used to capture and compress still images as digital files. The files are then controlled by ActionMedia 750 Software Library through program code written in C, or through an authoring language.

The analog source material, which can be photographs, slides, or videotape, are converted into a DVI® format by VCapt. This creates a still-image file.

VImCvt is the tool used to convert image files from one format to another, and to crop (take a portion of it for display and throw the rest out), scale (resize), or convert between DVI® and other still-image formats. For example, if you were interested in using images from a multimedia application you created on a Macintosh computer, or with another PC graphics card like the Targa board, you would use VImCvt to change each image so that it would exist in DVI® format, and could be displayed on a PC with ActionMedia hardware. You can think of VImCvt as a language translator for still images.

VShow is the tool used to display images. VShow can be used during the production process to display any image you have in a DVI® format. This allows you to look at each image, and make decisions about whether it needs to be edited or changed in some way prior to use in the application.

As covered in Chaps. 2 and 5, the ActionMedia 750 Software Library and the Production Tools are independent software packages. The Software Library is a complete multimedia environment. It is a critical piece of DVI® applications, since it manages all the multimedia elements when it is being played or used. The programmer uses this library to write the code that controls how, when, and under what circumstances the elements are displayed.

CAPTURING STILL IMAGES

How does a still image "get into" the computer? Still images are digitized and compressed using the DVI® workstation. The images can come from 35mm slides, transparencies, or hard copy (photographs). In most instances, you will use another computer peripheral, a digital scanner hooked to the computer, that "reads" the image, converts it to an RGB signal, and stores it as a file on the DVI® workstation's hard disk. You can also capture still images from videotape, videodisc, or from a camera source. This method will require that you add an analog-to-digital (A/D) converter to the DVI® worksta-

WHAT ARE DVI® FORMATS FOR STILL IMAGES?

Each software program that you buy for a personal computer has an underlying protocol, transparent (yet critical) to you, that allows it to be shown on the computer that you own. When you purchase software, you immediately look to make sure that you are buying it for an IBM or IBM-compatible personal computer, a Macintosh, Amiga, or other platform. You don't expect that this software will be interchangeable.

The same principle applies to still images and multimedia software. While there is work going on to change this (and DVI®'s underlying architecture is a perfect fit for these changes), the state of the art today is that images that are in one format cannot be displayed by another technology solution. That means that images that are in AVC format (IBM's Audio Video Connection), or CD-I's format, or the PICT format used in Macintosh computers cannot be shown on a DVI® system without conversion of the file format.

The fact that DVI® images are compressed is the most apparent reason for this. DVI® still images are stored using a compression scheme that reduces the amount of memory needed to store each image. They are decompressed and displayed by the i750 video processor used on the ActionMedia board. This compression/decompression scheme cannot be used by other computers—unless the i750 components are also used.

Work is currently underway to standardize compression and decompression algorithms for still images and motion video, so that all vendors will allow developers to use the same elements across platforms. One format is known as JPEG, for *Joint Photographic Expert Group,* for still images. The i750 components are able to display images in the JPEG format, and we anticipate that this capability will become more important in applications in a few years, when JPEG is used more widely. For more information about this proposed standard, and the one for motion video, see Chap. 17.

tion, since the ActionMedia board accepts an RGB signal. The analog signal from videotape, videodisc, or camera is an NTSC signal. (Some cameras can send an RGB signal. In this case the A/D converter would not be needed.) In general, this method does not produce as sharp an image as a scanner. For the highest quality, we recommend using a scanner.

More detail about these peripherals and how they are connected to the ActionMedia hardware is provided in Chap. 5. You will find that the performance (and price) of each peripheral varies. Some are best for slides, some for photographs. The methods and equipment used will have an effect on the final quality of the images in your application. We have made specific recommendations in Chap. 5 whenever we have a strong preference.

There are basically three ways you can capture still images for use in a DVI® application:

Lumena for DVI® technology and the ActionMedia boards

Targa board and associated software

ActionMedia 750 Production Tools and ActionMedia boards.

The peripheral equipment you choose will probably determine what method you use to capture still images. If you choose a scanner, it must have software that allows it to the "talk to" the hardware and software you are using. VCapt, the ActionMedia Production Tool software for capture, allows camera input only, while Lumena for DVI® technology has a driver to accept information from scanners. (One exception to this is a Sony scanner that operates as a camera, and can accept still images right into the VCapt software.)

Lumena for DVI® technology will accept input from a slide scanner and from a flatbed scanner (used for photographs). Most scanners will also be able to communicate with a Targa board. Images digitized through the Targa board can easily be converted to DVI® format with VImCvt, a part of the ActionMedia Production Tools. This is very useful if you are working with an artist who has a Targa board and is familiar with it, or if you have existing equipment that can be used by an artist while other team members use the ActionMedia system.

Slides can be digitized using a scanner that is made specifically for slides, or by attaching a slide scanner option to the flatbed scanner for photographs. A high-end slide scanner (about $10,000) will create very good-looking images. Images from slides tend to have a high degree of luminance and crispness. Slides must be very clean when they are scanned, however. A speck of dust can end up looking like a small crater on your image.

Slides and prints can also be captured directly using VCapt. Any RGB camera can use VCapt as the software tool for capture. There are also copy stands available (see Chap. 5) that will allow you to capture from slides or photographs. These copystands are equipped with an RGB camera, lights, and adjustable platforms to place the image.

STILL IMAGES FOR DVI® TECHNOLOGY

The production options for digitizing images is one variable to consider when planning your multimedia application. Another important decision that you will need to make is what image format to use. Options for image formats in DVI® technology give developers a maximum of flexibility. Still images can be used from a variety of electronic sources, as well as from print material, photographs, transparencies and slides.

The choices made about what images to use, and in what format, will have an effect on the overall look of the application. In addition, these choices will have an impact on how quickly images can load into memory and be displayed, and on the amount of memory required for each application.

An understanding of DVI®'s two main capture and display formats for stills, 9-bit and 16-bit, is a valuable place to begin to understand the balance you will need to strike between format, interactivity, and quality. The two formats both produce compressed, digital images. Whenever you use a compressed format, you gain storage, but lose some detail in the image. The two formats sacrifice different types of data. Basically, the 9-bit format sacrifices color detail, while the 16-bit format sacrifices some color depth.

Still images on the personal computer

An uncompressed, all-digital still has both spatial resolution and color depth. Both contribute to the reproduction of any image, whether it is motion video, graphics, or still photographs on the computer screen. The spatial resolution is basically the X and Y axis of the display, and designates how many pixels are present on the screen. Typically, a higher spatial resolution provides a sharper, more detailed image.

Color depth is used in a digital environment, such as the personal computer, to designate a third dimension of an image: the amount of color information contained in each of the pixels. The color in an image displayed on a computer screen comes from a combination of red, blue, and green. (Unlike your paints in kindergarten, where all color came from red, yellow, and blue, the color displayed on a computer is a mixture of light, rather than a reflection of light.) Most color displays on a computer are capable of a 24-bit deep image. Each color is assigned eight bits of data for its display. When displayed at a spatial resolution of at least 512 × 480, a photographic quality image is the result.

DVI® technology still images

DVI® technology uses these same principles of spatial and color resolution in the compression and display of still images. DVI® still images use a format known as YVU display. Y is luminance (or brightness) information, while V and U are color information. Most compression schemes used for any multimedia technology are using YVU as a basis of the compression. This format corresponds to the way that television technology works, and borrows from important research done for the development of color television. This research tells us the human eye is much more sensitive to luminance, or brightness, than to color information. When early television engineers had to find a way to transmit color information along with the earlier black-and-white (luminance only) information, the YVU scheme was invented.

This scheme is used to display still images, and to compress them so that they can be stored efficiently for use on the personal computer. An image that is 512 × 480 resolution and 24 bits deep requires about 740K bytes of memory. A CD-ROM, with 650 megabytes of memory, could only hold 900 of these images. Still-image compression is a technique that can be used on personal computers without DVI® compression. Most multimedia systems, like IBM's AVC, use some amount of compression.

The advantage with DVI® is that the i750 video processor is a special purpose, high-speed component that can compress and decompress still and motion video images in real time. Images, even compressed images, on a personal computer without the i750 video processor would have to slowly paint the screen with each image, and could not achieve the high compression rates with the high quality of images using DVI® compression.

Still-image compression can be either lossless compression, where no original information is lost from the image, or lossy compression, where information is actually discarded as a part of the compression process. Lossless compression results in an image that requires more memory to store, and is not required for most applications. Art applications, medical applications, and applications with detailed maps or other detailed images may require that some images be compressed with a lossless compression algorithm.

Lossless compression achieves about a 2:1 compression ratio. That is, images are compressed to ½ their uncompressed size. Lossy compression achieves a much higher ratio. Lossy compression averages about 12:1 with excellent image quality.

Both lossless and lossy compression for still images are available in DVI® technology. The two formats for compressed DVI® still images, 9-bit and 16-bit, represent two different approaches to still-image compression.

The 9-bit format

The 9-bit format is a compression format that actually reduces the U and V (color) planes of information saved for any particular image. The color values of pixels in a 4 × 4 block are averaged, as illustrated in Fig. 8.4 (repeated here as Fig. 10.1). All the luminance (Y) information is maintained to display the detail in an image. So, for every eight bits of luminance information, there is one bit of color information. Thus, the name *9-bit mode.* This is essentially a spatial compression scheme. The depth of the color information is not affected, but

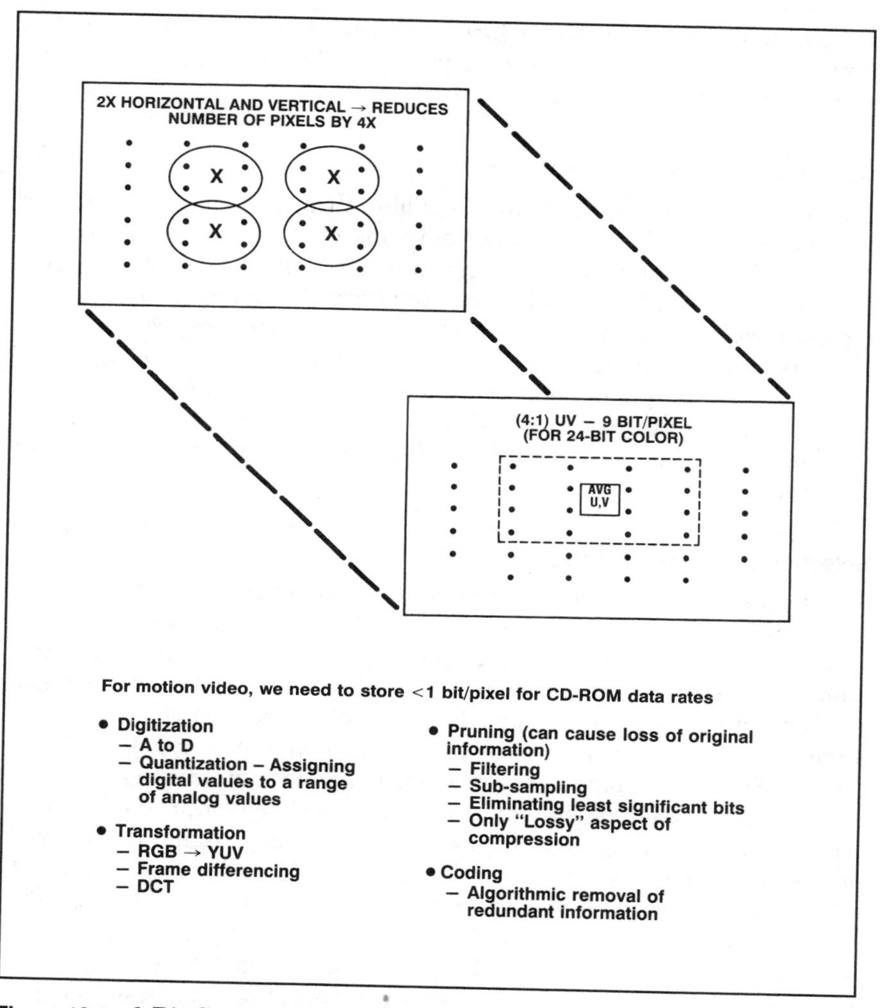

Figure 10.1 9-Bit Compression for PLV.

the color is averaged over the spatial resolution of the screen display. Since all the information in an image is not retained, this format falls into the lossy compression category.

The 16-bit format

The other major format available is 16-bit mode. In this format, color depth is averaged, rather than spatial color information. This format has full spatial resolution, then, but has less color information. The color information is not cut off, but is averaged among the 10 bits that are available for color. 16-bit format can be done as lossless or lossy compression. Lossy compression yields a higher compression ratio and lower file storage size.

Conversion between formats

The 9-bit and 16-bit formats are available through VCapt, the capture tool. Once an image is captured using one of these formats, it can be converted to other formats using another tool in the ActionMedia Production Tool software, VImCvt. However, the data that is lost during compression into 9-bit or 16-bit formats is never recovered, so for best results, you should capture the images in the format you plan to use in the final application. Another alternative is to capture all the images you plan to use in a 24-bit, or a 16-bit uncompressed format, and to convert them into the desired format later in the development process.

Selecting a still-image format

The theory behind the compression of these two formats hints at some important production implications. There are some memory and load-time implications—the less memory an image requires, the less time it will take to load, decompress, and be displayed. But the major production implication is how the application, and each image, will appear on the screen. For some applications, you may actually change formats from one image to the next, mixing and matching as needed.

The 9-bit format is the same format used for compressed motion video. If a still image leads into a motion-video segment, you will probably want to use 9-bit mode. Sharp, abrupt changes in format, which you would see if you were moving from a 16-bit still to a 9-bit motion video sequence, are slightly noticeable by the eye. However, if you are using text or sharp thin lines in an image—text overlay, for example—the 9-bit mode may pose a problem. Consider the fact that for every one pixel, a total of six adjacent pixels will change color value slightly because color values are being averaged over a 4×4 area. While this works well for video images, sharp graphics with one-pixel-wide lines could blur depending on the color contrast.

In 16-bit format, smooth areas of an image, if they span a large space, can show an effect known as contouring. This is because subtle differences in color depth are lost as the color values are averaged (or quantized) in the image. Yet, no spatial resolution problems, or bleeding, will occur. These images are larger in size than the 9-bit format images. Lossless and lossy compression are available in 16-bit format, while the algorithm for 9-bit images produces only lossy compression.

One important note about selecting an image format for capture: If you are not sure what format you want, it is best to capture in 9-bit

PRODUCTION TIPS FOR 9-BIT IMAGES

The 9-bit format is desirable because images are compressed to a very small file size, and because they are the same format used in motion-video compression. As with any lossy compression scheme, however, some information is lost. We have found the following techniques useful in planning a production with 9-bit still images.

- Use 9-bit still images to transition into a motion video sequence. Compressed motion video uses 9-bit as a display mode, so the transition will be smoother to the eye.
- When using text overlay, use a font larger than 12-point. This will help reduce the effect of pixel averaging that makes smaller fonts difficult to read.
- White text on a dark background is the best way to show text in 9-bit format. Design your stills so that the text always occurs in a bar, or on a button created in Lumena for DVI® technology. Colors for the bar are best black or gray with white letters.
- Avoid red and other deep, saturated color, as borders, graphic, or text overlay.
- Do not use single-width pixel lines, horizontally or vertically. These lines will blend with surrounding pixel colors, and will not appear sharp. (16-bit format is a much better production choice for an image with sharp lines.) Because the 9-bit format averages pixels in groups of four, the best design of any graphic overlay is to end its display on a multiple of four.
- A special 9-bit mode, known as *9Y,* is also available. This format reserves one bit of luminance (Y) for graphics. If you want to combine graphics on motion video or on 9-bit stills, you can use this format with almost no difference in the image.
- Audio files do not play over 16-bit to 9-bit transitions (or back) without an extensive programming effort. Evaluate the need for audio carefully, and break audio files up into smaller files, if needed.

and convert to 16-bit. That way, you maintain all the color depth and expand the spatial resolution in the conversion. By capturing in 16-bit, you lose color depth, then you lose spatial resolution in the conversion—the worst of both formats.

Selecting an image resolution

Another significant variable that will affect the quality and size of still images in your application is still-image resolution. The ActionMedia delivery board can display still images at varying resolutions up to 768 × 480. VGA still-image resolutions can also be displayed through the board. The absolute best way for you to choose a resolution and still-image format is to test a few of the images you will be using, and to make a judgment yourself, or by through user testing.

The experiences of other developers may help you with some basic guidelines, however. Are there noticeable differences between resolutions? Our recommendation, for most applications, is for still images to be captured and displayed at about 512 × 480 (or the equivalent VGA resolution). This is based on our own preferences, along with some preliminary research results from DVI® application developers.

Both the Getty Museum of Art and Virginia Polytechnic Institute and State University have conducted tests to determine if there is a threshold of display resolution for still images that is perceived by the human eye. The studies were conducted by showing similar images in pairs, and asking the subjects to choose a preferred image. The results are similar. Both studies found that people generally cannot distinguish the differences between 512 × 480 and higher-resolution images.

Granted, display technology for the two studies differed, as did image technology; the Virginia study used DVI® images and display, while the Getty experiment used a high-resolution workstation capable of displaying images up to 2000 pixels × 2000 pixels. Likewise, the audiences were different. Virginia Polytechnic Institute and State University used students, and the Getty Museum used art historians and museum curators. But the results are strikingly similar, and back up our own subjective view of this question.

VGA display of still images

The ActionMedia delivery board—the board used for playback of DVI® technology applications—has two display modes. DVI® output is interlaced as an RGB display at NTSC (15.75 KHz × 30 Hz) scan rates. This typically causes some flicker to occur on images with detailed one-pixel-wide lines.

DVI® output can also be displayed on a VGA monitor at VGA scan rates (31.5 KHz × 60 Hz). The noninterlaced VGA display mode eliminates the flickering. The maximum display resolution possible when you are using the VGA display for still images is 512 × 480. This capability can also be used to display VGA graphics over DVI® images, and to place video on a VGA graphics screen by creating a transparent "window" on the screen where motion video can be positioned for display.

THE IMAGE CAPTURE TOOL

Since the software tool that allows you to capture and manipulate still images is a part of the ActionMedia 750 Production Tools, you can capture still images right at your DVI® workstation. The tool is VCapt. It is set up so that you can capture one still image, customizing each one, or you can capture a series of still images with the same variables for each.

The menu that appears on the monitor of your DVI® workstation when using VCapt is shown in Fig. 10.2.

Navigation through this menu is very easy. The up and down arrow keys on the PC keyboard are used to select resolution. The right and left arrow keys are used to select either the 9-bit or 16-bit format. A filter level is selected by pressing the page up or page down keys. This software filter gives you results very similar to placing a filter over your lens in a camera. Filters soften the look of a picture. (Using a filter setting of "fair" is our recommendation for capturing images.) Selections are highlighted on the menu.

The image is saved by striking "S" on the keyboard. A prompt for a file name for the image then appears. Extensions will automatically be supplied to the image name, based on the selected format. If you name a file "boat" for example, and you have captured it in 16-bit format, it will be saved as "boat.i16". Selecting "C" returns you to the still-image capture menu. This selection also allows you to toggle between the live video and the still frame that is captured. You can use this feature to check the quality of the image capture.

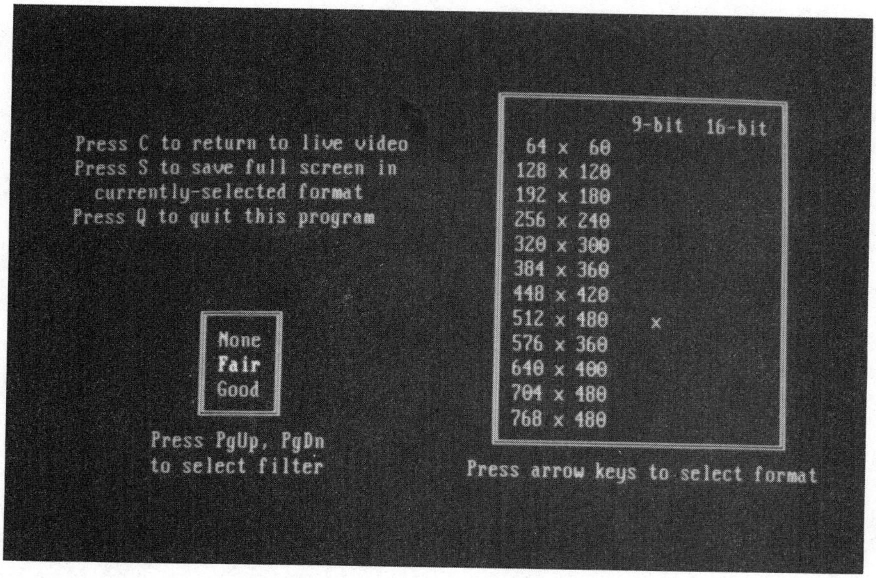

Figure 10.2 Menu for VCapt Production Tool.

You can also capture images in an automated, or noninteractive, mode using VCapt. By entering an option in the command line, VCapt will default to a command line interface, allowing you to capture images without the interactive menu. At the command line, you can control for format, resolution, filter level, and can select to have VCapt print a message and wait for a key stroke prior to capturing. The default for all of these options is:

Format 9-bit

Fair filter

512 × 480 resolution

No pause before capturing

An extension will, again, be added to the name of the file automatically.

The advantage of using this method is that a number of images can be captured in succession, using default, or by quickly using options at the command line without interacting with the menu each time.

An example of the command line option is:

```
VCAPT-B16-FN-P IMAGE1
```

This batch command will capture a still in:

16-bit format

Using no filter

Using the default resolution (512 × 480 as none is specified)

After a pause for a key press

As image1.i16

By using the F3 key to repeat a DOS command, you can work very quickly using this method.

THE IMAGE CONVERSION TOOL

After an image is captured as a DVI® image file, you may want to change the display formats. Or, as we mentioned earlier, images that have been previously captured in other formats can be imported to DVI® format. The tool used for both of these functions is VImCvt.

VImCvt is used at a command line to identify the input image, the input image format, output image format, and the name of the output file, if needed. Let's look at an example:

```
VIMCVT-F IMAGE1 9 C9 NEW IMAGE1
```

This command takes an uncompressed 9-bit image, and converts it to a compressed 9-bit image, changing its name to new image1.

VImCvt supports all of the following image formats. This list includes DVI® image formats, as well as some other common formats for still images.

9-bit format

8-bit CLUT format

16-bit format

24-bit format

Compressed 9-bit format

Compressed 16-bit format

Compressed 9-bit in AVSS format

.PIX—Lumena image file format

.TGA 16-bit Targa image file format

.TGA 32-bit Targa image file format

Raw 24-bit data format

Of course, as discussed earlier, different images will look very different in various formats. Essentially, each format has strengths and weaknesses.

Keep in mind that the maximum resolution for any image that is being imported or converted to one of the formats supported by DVI® is 768 × 480, the maximum resolution of the ActionMedia products today. Images being converted from Lumena or from Targa format must be 512 × 480 or less.

VImCvt options

This same software tool, VImCvt, also has options that will allow you to scale (size) an image, or to crop (cut out a specified region) an image. Some of the most relevant options, available at the command line, are:

-c Allows cropping, parameters for length of X and Y are needed, as well as an optional X,Y starting point. (If the image is in 9-bit format, the images should be cropped in multiples of 4.)

-s Sets the size of the output image, parameters for resolution X and Y needed. Must be 768 × 480 or smaller. Used in combination with -c, this option sets image size after cropping.

There are more VImCvt options available. All are covered in the ActionMedia 750 Production Tools documentation.

IMAGE SIZE, MEMORY, AND LOAD TIMES

The 9-bit compressed format is a desirable option for application development because it is the format that requires the least memory. A sample image, captured at a resolution of 512 × 480, will require the following memory in the following formats:

8 Bit/Pixel CLUT	240KB
16 Bit/Pixel Compressed	120KB
9 Bit/Pixel Compressed	60KB

These numbers affect the amount of storage required for the overall application; they also affect load times from the storage device, be it CD-ROM, local hard disk, or LAN server. The smaller the image (in memory size), the less time it will take to be displayed.

BUDGET, PRODUCTION TIME LINES, AND STILL IMAGES

The decision to use stills as a way to communicate in multimedia is mostly a content decision. It is also a budget decision. There are production advantages to the level of control you have over how images are captured and displayed. You can change them, enhance them, crop and scale them, and use them in conjunction with a paint program to create new graphics. You can use prints, slides, or camera as input, and you have a lot of control over the format and quality of your result.

Other than the budget needed for acquisition of still images and for a photographer, though, it is also useful to understand how the use of stills will impact your budget in production. Equipment costs can vary, as we have already discussed. In addition, you will need to budget time and resources for gathering images, digitizing them, and converting them as needed. Your Excel budget example in Chap. 6 has cells for each of these costs.

This sample spreadsheet also provides estimates for budgeting personnel time to digitize images and to convert them. We estimate that each image will take about 15 minutes of personnel time to scan, convert, and save.

Still images are a powerful communication tool. Many still-image libraries are available to use in multimedia applications. Some of these are public domain images from the government or NASA. Other libraries are becoming available as clip-art for multimedia. Combined with audio, special effects, and overlay graphics, still images take on a new life. The options for still-image capture, compression, conversion, and display help make still images a flexible and affordable media element.

Chapter

Producing Full-Motion Video for DVI® Applications

Compression of full-motion video is one of the central concepts and technical implementations behind DVI® technology. We have mentioned the theory of compression of motion video in several places throughout this book, and will review it once again here. The main purpose of this chapter, however, is to point out the implications of compression for production of motion video. In addition, we will cover some basics about how motion-video files can be manipulated and edited on the DVI® system during application development.

Review of compression principles

The essential function of compression is to significantly reduce the amount of digital information required for full-motion video, so that it can be economically stored and played back through standard personal-computer storage devices like CD-ROM, hard disk, or LANs. Compression is only one-half of the process involved in making this possible. Decompression is the other half. The processes of compression and decompression are also referred to as *encoding* and *decoding*.

Regardless of nomenclature, the analog signal produced by video must be converted into digital format prior to compression. This process is referred to as *capture*. Compression is an algorithmic reduction in the amount of information contained in each frame of video. Decompression involves retrieving this information and reconstructing it for display on a computer monitor. Compression can be done either in real time on a PC equipped with the appropriate DVI® hardware and software (RTV), or can be done in an off-line process in a compression facility for the highest quality (PLV). Decompression is a real-time process, initiated by a command to the PC through the DVI® hardware and software.

Although compression of motion video is a complicated, sophisticated process, in essence, off-line compression (PLV) is achieved by comparing frames within a video sequence, and storing the differences. This is covered more extensively in Chaps. 2 and 8. Obviously, given this as background information, the more new information introduced in any single frame, either through fast motion, the introduction of new objects, or the extremely fast movement of a camera through a scene, the more difficult the compression task becomes. This is one production consideration introduced by all-digital multimedia.

PRODUCTION PLANNING

The time to begin thinking about how you should produce motion video for your application, or what type of stock footage you should select, begins early in the design phase. The two broad areas that a video producer should consider in creating video for multimedia are *design* and *technical.*

Design implications

In most ways, producing an interactive, multimedia application is very similar to producing a standard linear video program. A good script is essential; high-quality production values will be most effective; good on-camera performances will capture the audience's attention. Therefore, regardless of whether PLV or RTV compression techniques will be used in an application, a video producer should not abandon the principles of production that would be important to any production.

At the same time, most video producers will need the expertise of a good interactive designer to help shape the production. This is a new area for most video producers. Full-motion video, in contrast to a linear videotape production, is only one element of a multimedia production. The interactivity, not the linear story, is the central feature of the multimedia application. The truly effective multimedia producer uses all media sources with an eye for creativity, storage space, and effective interactive design. With this in mind, the following guidelines may be useful:

- Avoid extensive use of "talking head"—video sequences that feature on-camera talent talking directly to the camera. We have found that introducing a character, and establishing the voice and interaction with the user, is a good use of this type of video, but that instruction or lengthy technical data is not effective delivered in this way.
- Use full-motion video for short, strong sequences. Anything longer than three minutes should be evaluated carefully.
- Any motion-video sequence should have a goal associated with it, as any media element you use. This will help focus the production team on what specifically should be the point of any particular shot.

No rule of thumb can answer the question, "When do I use motion video?" Before planning a full-motion video production, ask the following questions:

1. Could this information be presented as effectively with graphics, stills, and audio?
2. Can this segment be broken down into smaller, shorter pieces to be more effective?
3. Is there a distinct advantage to incorporating interactive responses, or branches, to break up long, sequential video pieces?

Interactive applications should have different goals than linear programming. A video producer may need experience or assistance as these goals are developed, and as production is completed.

Technical implications

Some of the considerations that fall under the technical heading are familiar to the video producer. Selecting a format for the collection of images (film, video, etc.) is a part of any production, as is creating good lighting, etc. Other considerations are specifically introduced by the all-digital format. In the technical production area, some information will apply specifically to PLV or RTV compression formats as well.

Video formats

Video can be recorded in a variety of tape formats. These formats range in quality from VHS, which is used almost exclusively for home/personal recordings, to 1″ video tape, the high-end broadcast-quality video format. (New digital formats, like D2, may in fact replace 1″ tape in the future.) Along the continuum, formats such as ¾″ video tape, BetaCam, MII and other recording formats, also are available. There are advantages and disadvantages to each of these formats. For the most part, the more expensive the format and the recording equipment that goes with it, the higher quality your resulting motion-video imagery will be.

Currently, compression facilities for PLV only accept the highest quality, 1″ broadcast tape. In the future, these facilities will probably diversify and specialize in different formats. If you produce video on any other format, it can be "bumped" or duplicated to 1″ tape in a video editing suite.

You can achieve good results by bumping from BetaCam SP to 1″ tape, and we recommend shooting on BetaCam SP for this reason, along with the fact that it is much more economical that shooting on 1″ tape. Bumping from ½″ VHS to 1″ tape is not recommended, however. In fact, if your production does not require the quality of Beta SP and 1″ tape production, you may not benefit from PLV compression at all. *You may want to consider RTV compression for applications that do not have a need for the highest-quality compression.*

An important note: *Never,* never submit your camera originals or your original edited tape for compression. You should always work from a protection copy, and keep your masters in safe storage.

Shooting video—the traditional guidelines

Traditional video guidelines could fill a book. Advances in production technology have reduced the complexity of production, but it is still an art/science that requires a significant knowledge and continuing education. New techniques often violate old rules of production. For example, random camera movement, once taboo among producers, is now an accepted technique. The advent of chip cameras has reduced the problem of shooting video in low-light situations, and in high-contrast lighting such as direct sunlight, bright flames, and surfaces covered with reflection. Even color saturation, a production error that once caused bright colors to "bleed" onto each other, or flicker dramatically, can be corrected in postproduction. Traditional video guidelines are a set of rules that are changing. A producer must look at the equipment available, the desired outcome, and the content being conveyed to determine the do's and don'ts.

When the exact technical parameters are determined, it is a good idea in any video shoot to stay on the conservative side. This is essential when shooting video that will be compressed. Compression usually compounds problems associated with video artifacts. Moiré (an effect where small lines—like those on a pin-striped shirt—seem to "dance" on the screen), poor lighting, color saturation, quick or random camera movements, and other problems in a traditional shooting session, are potential sources of trouble for the compression process, as well. Compression does not correct video artifacts; while it is a digital process, it should not be confused with digital postproduction. The compression algorithms merely read the information in the frame. They do not automatically correct for color, lighting, or other poor production results.

These effects will be visible in both RTV compression and PLV compression, though maybe more problematic in applications where higher-quality PLV compression is a requirement.

Camera movement

Since the PLV algorithms are based on saving changes between frames, introducing quick, random movement is not recommended unless it is absolutely necessary in your application. Instead, use smooth, subtle camera movements. If movement is needed, one alternative is to use a dolly, preferably on rails that will stabilize the camera. Handheld camera work, as a rule, should be avoided, unless it is supported by some special equipment mount like a Steadicam. Long truck shots can also introduce vertical camera movement that will be problematic in compression. In general, avoid quick pans of ten frames or less. The RTV and PLV compression algorithms are based on different principles, and RTV does not use delta-encoding. Quick, random movements introduce more problems for production that will be used for PLV compression.

Camera filters

Many video producers use filters to soften the video picture. When shooting video that will be compressed, keep in mind that filtering is applied as a part of the compression process. If you would like to use a filter to achieve a softened picture, select a relatively soft filter. The resulting compressed video will look much like video shot with a medium filter. This is true for both PLV and RTV compression.

Video transitions and special effects

Transition effects that can be done in software by DVI® technology are for still images only today. (Some effects can be done in software on motion video at smaller than full-screen size or at video playing at 15 frames per second.) If you want to have special effects in your motion-video elements, you may want to add them in postproduction. Since these effects will be treated the same way the motion video is, these transitions should be done smoothly. Transitions of less than 15–20 frames in video targeted for PLV compression should be avoided. Transitions at the end of any clip, like a fade or cut, can be done in software by holding the last frame as a still image and applying the software transition.

Audio postproduction

Audio that is synched with the motion video should be mixed prior to submitting the video for compression. These files will be returned with both audio and video in synch. While it is possible to mix your soundtrack after compression, unless there is a design reason to do so, you are creating more work for your team. This process is automated in the compression facility; there is no extra charge for it, and it will save you time and effort.

Resolution of compressed video

Resolution refers to the number of pixels that make up a picture on the PC screen. Each pixel represents a "dot" on the monitor, and all images are displayed using this dot pattern. Most graphic artists, computer programmers, and others familiar with this notion also assume that the higher the resolution of the image, the higher the quality of the image. This is true in graphics, and in some cases still images, but pixel depth also plays an important part in the quality of images when displayed on the PC.

While the resolution of the display of motion video is 256 × 240, this cannot be correlated to the same resolution as it applies to graphics. Instead, you should think about this resolution as about VCR quality. You should also keep in mind that this motion video can be played back on a screen that is in a 512 × 480 resolution mode. It will play back as ¼ of the screen. The image has not changed resolution; it is just being displayed on a screen that has more pixels. The result is that it plays in a fourth of the available area, and it appears to be sharper.

Reference frames and delta frames

The compression scheme for PLV introduces the most significant technical difference between producing for analog video and compressed digital video. *Reference frames,* or *full frames,* are the frames in the video that represent all the video information. *Delta frames* are all the other frames in a video sequence. They are the frames that only hold the new information, or the information that is different from the one just prior to it. This concept is discussed in Chap. 2, and illustrated in Fig. 2.5.

Reference frames are automatically inserted at the beginning of each file you have compressed, at scene cuts, or in instances when a lot of new information appears on the screen at once. They can also be inserted at regular intervals at the compression facility if designated by you. You may want to do this to break a file up into small pieces. If you want to insert reference frames regularly, no more than one every 30 seconds is recommended. Of course, in RTV compression each frame holds all the information, and this consideration is not important.

Cropping video in compression

The compression facility for PLV does provide a service called *cropping* for customers, and cropping can be done on RTV files using the production tools. Because of the wide discrepancy in the actual viewing areas in camera viewfinders and monitors, video will sometimes contain unintended or unwanted objects in the frame. Eliminating this information by throwing out a specified edge of the frame is called cropping. You can specify cropping parameters in pixels to a PLV compression facility, ensuring that your video files contain only the information you want.

Decompression or decode time

The i750 video processor, as discussed in Chap. 2, is the workhorse that takes all of the compressed video frames and decompresses and displays them at 30 frames per second on the PC monitor. In some instances, the video processor is also displaying graphics at the same time. This work takes significant processing power. The PLV algorithms used in a compression facility were designed to match the power of the i750 video processor. There is a variable in the algorithm that can be set, however, to control how much of the processing power is spent on decoding the compressed motion video. This is called the *decode time,* and is specified as a value between 1.0 and 2.0, in increments of 1/10.

The decode time is important because it can be used by the developer to trade off image quality with computer-processing capability. A value of 2.0 means that 100 percent of the i750 video processor is dedicated to decoding and displaying motion video. This is hardly ever used, since graphics display, cursor, and other activity on the screen is assumed in an interactive application. The default used by compression facilities is a decode time of 1.9. This decode time is adequate to display static graphics on the screen with video.

For some applications, and some uses, you may want to have more processing power for scaling and resizing video, complex graphic displays, multiple video streams, or other digital manipulations. You can specify a lower decode time for video, and depending on the image, the difference in quality will vary. For a simple video scene with little motion, decode time may go as low as 1.4 with little effect. For complex scenes, we recommend 1.6 or higher.

RTV has a similar setting in the VRtv production tool. VRtv allows variable settings for file size. Obviously, the higher this setting, the higher the quality of image that will result. However, higher file size settings will take more processing power, with less left over for other operations.

You should determine what decode times or file size you need as a part of the initial design process, and in consultation with the interactive designer and software programmer or designer. These settings may affect how you shoot the video, and the video may affect what is required. This is another example of how a team must communicate and work together to make a successful multimedia application.

HOW TO ORDER COMPRESSION OF PLV MOTION VIDEO

PLV video compression is done through an Intel-licensed compression facility. You will send a 1″ videotape of the material you wish to have compressed, with an order form and purchase order, and you will receive the compressed video files back on a high-density data cartridge. In order to avoid errors in the compression process, and to ensure quality, Intel has developed standards and procedures for the video as it arrives at the compression facility.

Compression facilities keep things closely scheduled to best utilize the equipment used in compression. Like any off-line postproduction process, video compression must be scheduled ahead of time. We recommend that you schedule your PLV compression no less than two weeks prior to sending your tape to the facility.

As with all professional video processes, time code is used in the compression facility to assign readable addresses to each frame of a video sequence. SMPTE time code, the NTSC standard in North America, is used by almost all video production and postproduction facilities, and licensed compression facilities use it as well. All analog tapes submitted for compression must have continuous, contiguous nondrop frame, address track time code. The forms submitted with video, available from the compression facility, provide space for each video sequence to be identified by a file name and a start and stop point designated by time code, as illustrated in Fig. 11.1.

When your videotape arrives at a compression facility, it is viewed by compression service engineers to ensure that it is the correct tape, to check that the time codes correspond to your request, and to check for overall quality. If problems are found, a compression services engineer will call you for clarification. If not, the compression process will begin, and you will receive a digital streamer tape with your files as scheduled.

intel®

Intel Princeton Operation
Compression Services Request Form

JOB # ____________

CLIENT INFORMATION

NAME: ____________ DATE: ____________
STREET: ____________
CITY: ____________ STATE: ____________ ZIP: ____________
PHONE: (______) ____________ FAX: (______) ____________
CONTACT: ____________

MASTER TAPE(S)

______ NUMBER OF TAPES ______ STORE OR RETURN?
COMMENTS: ____________

PROGRAM TITLE

SHIP TO

COMPANY: ____________
STREET: ____________
CITY: ____________ STATE: ____________ ZIP: ____________
NAME: ____________ PHONE: ____________

REQUESTED RETURN DATE

PAYMENT:

☐ PURCHASE ORDER (Specify P.O. # ____________) ☐ COMPANY CHECK

SHIP VIA

☐ FEDERAL EXPRESS ☐ UPS OTHER *(specify):* ______

ACTION MEDIA™ 750 BOARDS? ☐ YES ☐ NO

1. ORIGINAL FORMAT	2. ORIGINAL FRAME RATE (fps)	3. PLAYBACK RATE (fps)	4. PLV VERSION (if known)	5. BYTES PER FRAME	6. DECODE NUMBER OF FIELD TIMES

7. CROPPING UPPER LEFT (x, y)	8. CROPPING SIZE (W x H)	9. INS. REF. FRAME EVERY ______ FRAME	10. STILLS Y/N	11. GRAPHICS MODE Y/N

12. SEGMENT TITLE	13. FILE NAME	14. SMPTE IN	15. SMPTE OUT	# OF FRAMES	16. AUDIO Y/N
		: : :	: : :		
		: : :	: : :		
		: : :	: : :		

White - *Compression Services* Yellow - *Customer Support* Pink - *Administration* Gold - *Customer*
PROCOMP 1

Intel Princeton Operation · CN 5325 · Princeton, NJ 08543

Figure 11.1 Compression Services Order Form.

PRODUCTION TOOLS FOR YOUR MOTION VIDEO FILES

Once your video has been compressed, the video sequences can be incorporated into your DVI® application. Several authoring systems have become available that allow you to complete the process of creating an application, and are covered in Chap. 13. In addition, there is a set of production tools available from Intel to use in playing and editing PLV and RTV motion-video files. More recently, a desktop video editor, D/Vision by Touchvision Systems, has become available to developers for use with RTV and audio files.

Keep in mind that the production value of the two types of compression, RTV and PLV, does differ. While we anticipate that the quality of RTV will improve over the next year and beyond, there will always be a difference between the quality—with PLV being the better quality option. RTV's place may be for internal applications, video mail, and for prototyping actual applications. You will want to test a small piece of video with your target audience (or buying client) prior to selecting the compression format you want to use.

Production tools

Production tools capture elements as either image files (for stills) or AVSS files. These tools are covered generally in Chap. 3, and are

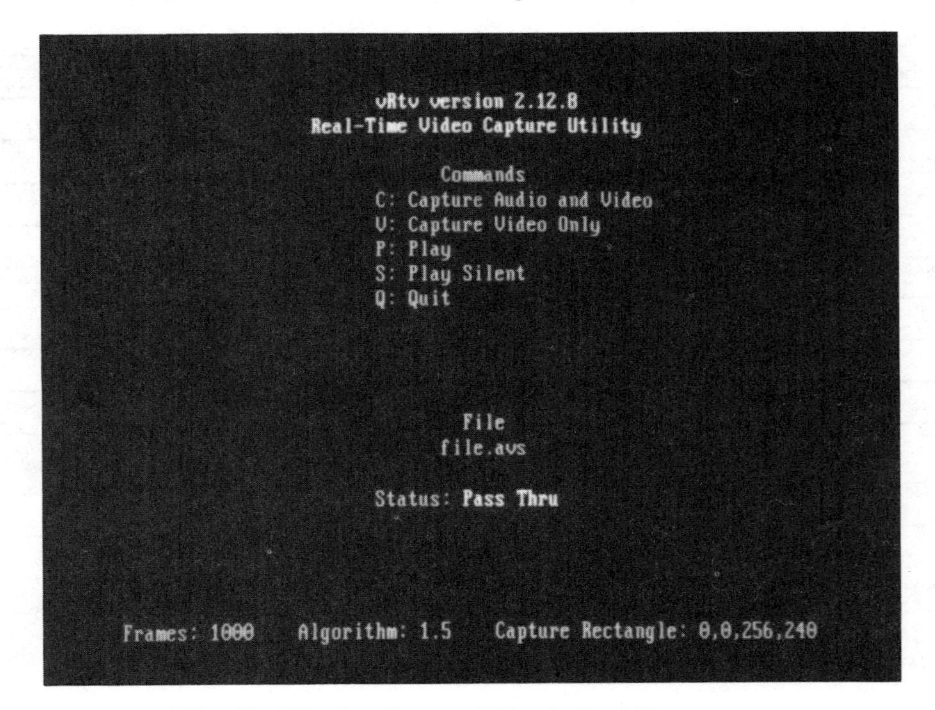

Figure 11.2 VRtv Tool Used to Capture Video in Real Time.

described here in more detail. VRtv is the tool used to capture video in real time, as illustrated here in Fig. 11.2. VAvEd is then used to edit these files, and VPlay is used to display them.

Steps to using VAvPrep and RTV

VRtv captures and compresses video with monaural audio if needed. VAvPrep is the first step needed in order to save an RTV file onto hard disk. VAvPrep is a "fast disk" preparation tool that prepares the hard disk so that the compressed motion video files can be played at 30 frames per second. VAvPrep must be used prior to playing PLV or RTV files.

VAvPrep is a command-line tool. It creates a control file that acts as a stand-in until you fill it with the RTV information you capture. After creating the control file, you can capture video.

The display resolution for RTV video is 256 × 240. The VRtv menu gives you the option to choose to capture audio and video or just video, and to play it back with audio or silently.

You can also capture video by using the command line, setting resolution, number of frames to capture, and an algorithm (1.0 or 1.5) as options. Defaults for this command line capture are:

256 × 240 resolution

90 frames

Version 1.5 (30 fps) algorithm

Changing the resolution at the command line will result in a cropped image, which is automatically centered.

RTV can also be displayed on a VGA monitor at VGA scan rates. The image is displayed in a mode that automatically doubles each line. This is called *interpolation*.

Using VPlay

The PLV compressed motion video files that are returned to you on digital tape are tested prior to shipping using VPlay, one of the tools included in the ActionMedia 750 Production Tools package described in Chap. 2. Using VPlay, you can verify that your PLV or RTV video plays from hard disk or other digital storage device. VPlay is a command-line call. After preparing your hard disk using VAvPrep, you simply type:

```
VPLAY FILENAME
```

for each file you want to view.

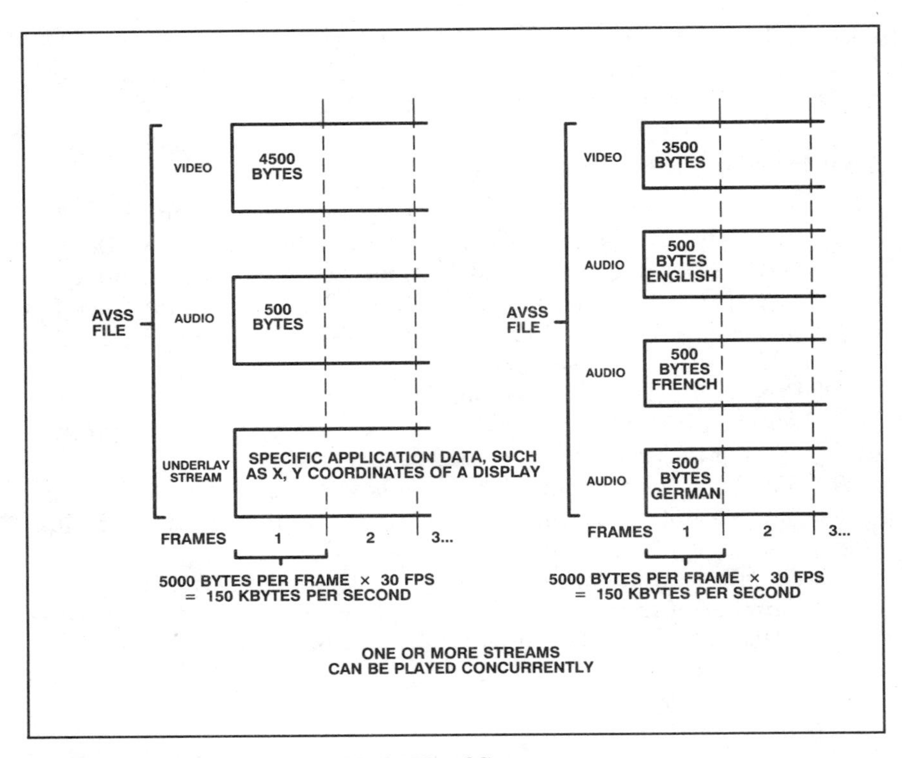

Figure 11.3 AVSS Files Are Made Up of Streams.

Editing motion-video files

VAvEd is the tool used for editing motion video. This tool can be used to shorten video sequences, edit sequences together, and to mix new audio with existing video files. You need to know three specifications with VAvEd. The first is the name of the AVSS file (PLV and RTV files are stored as AVSS files, as are audio files). The name of the file is simply the DOS name you have assigned.

An AVSS file is made up of streams. Typically, these streams are audio and video, as shown in Fig. 11.3. If you are using VAvEd to edit video, you must identify the stream of the file that contains the video you want to edit. Last, you will need to know the frame number or range of frame numbers you want to work with.

In the following example, VAvEd is used to create a new AVSS file, containing only frames 5–10 from the source file.

```
COPY [SOURCEFILE] (S 0) [DESTINATIONFILE] (S 0)
COPY [SOURCEFILE] (S 1) [DESTINATIONFILE] (S 1)
```

```
COPY [SOURCEFILE] (S 2) [DESTINATIONFILE] (S 2)
WRITE [DESTINATIONFILE] 5 10
QUIT
```

This set of commands has copied stream one and two, in this case audio and video, to the new file.

Other video editors

While VAvEd is a powerful set of commands for the personal computer, the interface is not one that video producers, or multimedia producers for that matter, are familiar with. New video editors are currently available for use, including D/Vision by Touchvision.

D/Vision is one of the most exciting new products for use on a PC. This software package is designed to be used with RTV files. It enables the multimedia producer to edit video clips for direct use in an application, or to specify the required segments for compression of PLV.

D/Vision uses a graphical user interface to approximate the controls of a common video appliance, the VCR. The creators of this application used proven editing software and hardware (used in the television and movie industries) as a model for the interface. Files are located, played forward or backward, "trimmed," and "lifted" to create new files that can then be played in a DVI® application.

Editing is precise, and achieved with frame-to-frame accuracy. Video material can be played back at variable speeds. Audio track editing can be done completely independently from the video editing. Edited material can be seen and heard immediately upon editing. D/Vision also accepts other files, like stills, animation, and titling, from other DVI® compatible software programs for editing. A sample screen from D/Vision shown in Fig. 11.4a and 11.4b (see color plate).

While this product is young at the time of this writing, early demonstrations and use by a limited number of developers show that it holds great promise as a low-cost, easy-to-use environment for editing video and audio. If you will be editing a great number of video files, you will want to invest in this software rather than using VAvEd.

The addition of motion video to application software for the personal computer makes the PC a complete communications machine. Text, stills, and graphics all have been used in applications to illustrate and impart important content. Now, another set of communication experts—those who have used motion video and audio—can use the personal computer as a way to send messages.

The addition of motion video widens the circle of knowledge and expertise needed on a multimedia development team. This is a new

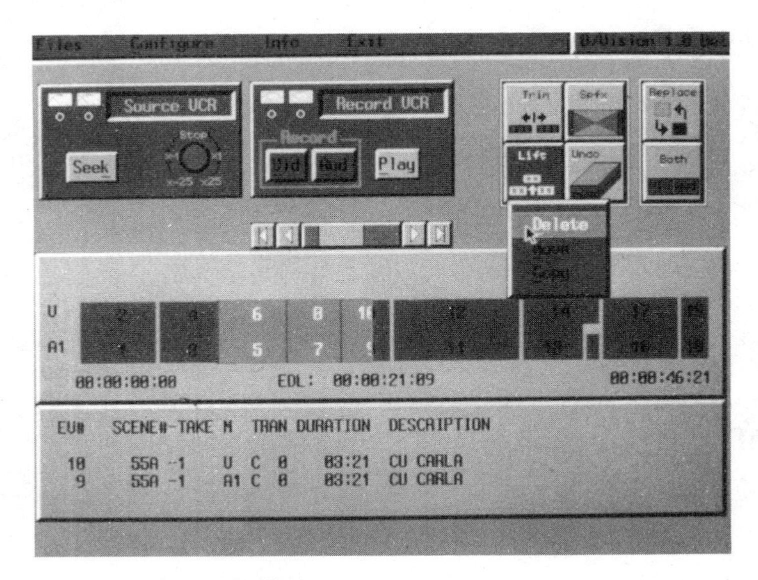

Figure 11.4a Interface to D/Vision.

discipline for software developers, albeit a fascinating one. Likewise, video producers may find the keyboard a bit disconcerting at first. Both the producer and developer, however, can find new power in the capability that DVI® technology brings to the personal computer. Good communication between these professionals will result in high-quality multimedia software for PC users.

Chapter

12

Creating Multimedia Soundtracks

The audio soundtrack is often not given the credit it deserves as one of the key elements in multimedia. Sound can bring pictures alive, and add depth and dimension to the visual content. Take a moment to visualize a picture of a forest which dissolves to a new image which includes a creek or brook. Now, play that back in your mind, but add the sounds of the forest. When our second image of the babbling brook is revealed, add to our imaginary soundtrack the sound of the brook. In our simple example, you can see how we mentally already start to make the audio link between images and sounds. In multimedia, we don't have to depend on our imagination. This capability and the tools to build soundtracks are available to us as multimedia producers.

Elements in a soundtrack

Like the graphics designer, the soundtrack producer has a number of options available to match the soundtrack to the needs of the project. When digitizing an audio soundtrack, various sampling rates are available that can be used to match the quality needs of the application. This concept is very similar to the ability to select the resolution and pixel depth for digitizing still images. Like image resolution, you can select the quality necessary for that portion of the application, and weigh it against the need for conserving data storage.

The soundtrack producer also has more conventional audio tools available to him or her. For example, stereo can effectively be used to create dimension or the feeling of motion. The effect of a car passing by, visualized with still-frame images, will be enhanced by panning the stereo sound from one side to the other.

Figure 12.1 shows the final frame of our opening visual build to our sample application called Horizons. The opening soundtrack was designed with audio effects mixed with the music to enhance the visuals as they transitioned one by one onto the screen. First, in the beginning was the sound of the ocean and the wind followed by seagulls flying overhead. Next, the chugging of an old gasoline engine was mixed to note the appearance of the Wright Flyer into the scene. Last, as the music began to build, the roar of a passing jetliner was added to complement the shadow of the jet in the center of the visual. The left-to-right pan position of the audio sound effects coincided with the appearance of the image.

This is an excellent example of how several small segments of audio sound effects can be built to complement a series of still frames. While still frames may not have the impact of motion video, adding audio to the presentation of a series of still frames can be as effective as motion video. Stills and audio can be very effective presentation

Figure 12.1 Opening Finished Frame to Horizons.

tools with a significant savings in file size and production cost versus digital motion video.

Audio sound tracks with digital motion video

When motion video is digitized and compressed with DVI® technology, the audio track can be compressed at the same time. This results in a single file with audio and video interleaved. When there is lip-synch video, the audio soundtrack must be accurately synchronized with the motion video, and interleaving the tracks in one file is the best way to achieve this.

If the audio soundtrack is digitized and played back separately from the digital motion-video file as two separate files, the application designer has the added flexibility of being able to interactively change the audio file based upon the user's interaction. This feature can be used well in soundtracks where lip-synch is not critical, like voice-over narration or music tracks.

For example, in a point-of-sale application, the primary digital audio file for a segment of motion video displaying a product may describe its features, advantages, and benefits formatted into a sales presentation for a customer. With careful design, the same digital video file with an alternate audio soundtrack could be used to present a training message for in-store personnel on special features of a product that should be pointed out by the salesperson.

Narration

Narration is used to explain a product or provide a training message to the visuals being presented. Narration can be done in several different languages, at different knowledge levels, or for other varying audiences.

Music can add a lively introduction and closing that sets the theme and tone for the application. When a presentation is designed to sell products, music plays a key role in setting the mood and tone the soundtrack producer desires for the customer.

Library or needle-drop music

There are a number of options available for using prerecorded music. Collections of library or needle-drop music can be purchased with the royalty fee paid on individual cuts, or as part of the one-time purchase price of a complete library. There are many different audio soundtrack libraries available, spanning a complete range of moods for use in developing a multimedia soundtrack, including classical music and upbeat contemporary music. The available selections cut across a broad range, including, for example, music to set the stage for various historical periods, such as a history lesson based on the Colonial Period of the United States.

A list of companies who distribute production library music is included in Appendix B. Library music is typically distributed on audio CDs to maintain the best quality. Audio CDs offer the additional advantages of being easy to cue during production.

Original music

Another option is to commission soundtrack music specifically composed for your production, just like scoring a movie. This used to be very expensive, as it often required a studio full of musicians to record a composition written by a composer.

Today, as a result of the development of electronic synthesizers and MIDI-based musical instruments, the composer using a personal computer and sequencer software can create a multilayered soundtrack that includes the sounds of the complete orchestra or band, all emanating from several electronic keyboard instruments and synthesizer systems.

The advantages to a custom-composed soundtrack are many. One, the soundtrack can be designed exactly to the needs of a particular production. The tempo and mood can be changed by the composer to match what's being displayed on the screen. The alternative with library or needle-drop music is the process of cutting together and cross-fading various music components to create the appearance of a custom-produced soundtrack. This requires some special skills and experience, although with today's user-friendly recording equipment, simple soundtracks can be easily produced.

Originally-composed music has a second advantage. The most popular collections of library music are often sold to radio and television stations and advertising agencies for use in commercials. It might not be unusual to find, for example, a local car dealership in the Chicago area using the same music used to open your multimedia training masterpiece.

Your decision whether to produce a soundtrack with library or needle-drop music will often depend on the budget available for the soundtrack production, its intended use and application, and the ability of the library music to meet your audio needs for your application.

Sound effects

In the opening to this chapter we illustrated the power of sound effects to take a simple series of visual elements and make them come alive. We took a series of still images and combined the music with the sounds of the wind, seagulls, and the motor from the Wright Flyer.

Sound effects for producing soundtracks like this are available from a variety of sources. Like library music, there are a number of companies that sell collections of sound effects on CD or tape. The sound effects are typically short audio clips organized with a printed directory. Some sound effects collections include thousands of sound clips. For example, take the sound effect for an explosion. Sounds simple, but do you want a sharp, short explosion like a firecracker, or the throaty, deep roar of what sounds like a chemical plant exploding? The sound effects collection listing under the category of explosion sound effects is extensive, to accommodate the needs of many different productions.

HOW DOES THE PROCESS OF SOUNDTRACK PRODUCTION BEGIN?

Creating an effective soundtrack begins early in the design stage. Depending on the skills and viewpoint of the creative team, some multimedia applications will begin with the visual material. Others many times begin with a script, including the narration and recommendations for the visual information. The best approach is probably to design the soundtrack in two passes. During the first step, the audio is designed together with the visual information. During the second pass, the audio soundtrack producer goes back and looks carefully at the finished production in storyboard form and with the recommended narration to determine where the audio can be additionally enhanced, either through the careful selection of music or the insertion of sound effects.

Like the visual information, during this portion of the application design development it is often a good idea to begin to identify the audio elements for the production as individual files through log sheets. Labeling these files during the design phase will simplify identifying the material later on. While you are creating your log of audio files, it is often a good idea to note other important parameters, such as the included sampling rate and the estimated file size based upon the length of the running time of the audio segment multiplied by the sampling rate. The file size is an important reference in order to keep your audio production within your application's megabyte budget. As stated earlier in this book, even though many applications will be distributed on CD-ROM with its capacity for 650 megabytes of data, it is surprising how quickly an application can fill this up. It is also important to track the size of files to determine what impact they will have on file transfer rate and load times.

Creating an audio time line

One of the best metaphors for designing an audio soundtrack is to create a time line. A time line can be created using either a graphics package or with a little programming, even using Microsoft's Excel program to visually describe in linear form the various visual and audio cues. Whatever format you elect to use, it is often a good idea to lay out the soundtrack against time as shown in Fig. 12.2. Visualizing the audio soundtrack against the probable images allows the soundtrack producer to determine when sound effects should be added or the tempo of music should be changed to match the visuals displayed on the screen.

It is often a good idea, once the soundtrack is plotted on paper, to review it with the programmer. The programmer can look at the soundtrack design to determine whether load times will become a factor, or the interaction of visual image transitions with key points within the soundtrack or the beginning or ending points during the playback of soundtrack files. As often stated throughout this book, there is never enough good planning to help avoid unplanned errors and mistakes. Note, in our example of the Horizon soundtrack, where we show the various points where still images are transitioned onscreen and their associated sound effects. Note also how this needs to be coordinated with the appropriate cue points in the narration.

Producing your own soundtrack

Based upon our multilayered example, the production decision is to produce this soundtrack in a recording studio, or by using one of the low-cost systems for home audio soundtrack production as shown in Fig. 12.3.

As a result of the development of low-cost MIDI-based synthesizer equipment, companies such as Tascam, Yamaha, and others have developed low-cost audio recording and mixing equipment for creating very good quality soundtracks, typically on high-quality cassette. Cost for these systems range from $500 for a four-track stereo mixing

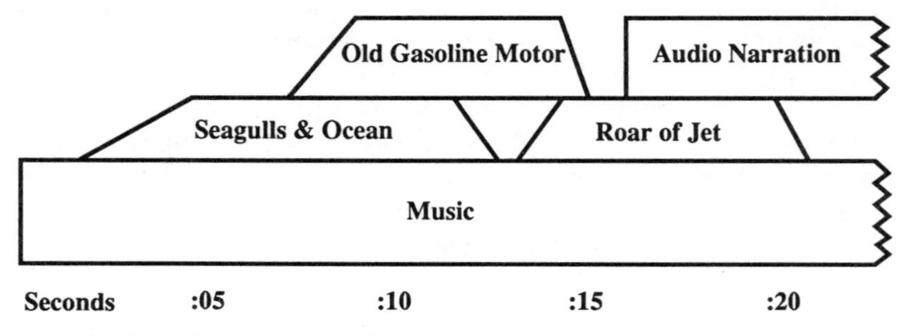

Figure 12.2 Sample Audio Time Line.

and recording system to systems allowing eight or sixteen tracks to be mixed to two or four tracks for a complicated soundtrack production.

If you do not own this equipment, or you find the needs for the special skills of an audio recording engineer, or access to a large collection of library music, you may elect to produce the soundtrack in a recording studio.

Using a recording studio

Recording studios can be found almost anywhere by checking the local yellow pages of a telephone directory. In small towns and cities, a radio station will often have a recording studio and engineers who also produce soundtracks for commercials.

The rates for soundtrack production can vary based upon the location, the capability of the audio engineer, and the audio recording facilities. Rates can range from $50 per hour to over $300 per hour. In many cases, when the design of the soundtrack is properly communicated, the recording studio will produce a soundtrack for a fixed price bid.

Narration—voice-over talent

One of the most important elements to set the tone and mood for your presentation is the selection of the soundtrack voice-over talent. Like

Figure 12.3 Low-cost Audio Production System.

selecting a recording studio, there are many people available, and often if you have elected to retain the services of a recording studio, they can be helpful in selecting voice-over talent for your production.

People who regularly do voice-overs often have demo tapes available that include a collection of sample readings of scripts. The readings many times vary showing different inflections and styles for different productions. For example, it is not unusual to find many voice-over talents who can provide the voice-over for character actors. This might be used, for example, in a historical drama type of application or in a training application when voices are used to act out a situation visualized in a series of still images.

In our sample application, Horizons, the style of the narration was designed to give it a historical perspective. The voice-over talent's voice and reading was also intended to add drama to each step in the evolution of manned flight.

The costs for using voice-over talent vary depending upon the popularity of a voice-over talent's voice and the application for the narration. The most popular voice-over talent, who are often used in national advertisement campaigns, may charge thousands of dollars for their work reading the script for your production. Many narrators and voice-over talent are available for fees of less than $100, however.

The variation in fees will depend also many times on the extent of distribution and the application. For a multimedia application that will be published and widely distributed, voice-over talent will command a larger fee for the use of their voice. This is one of the many creative and budgetary decisions facing the soundtrack producer.

Directing the soundtrack production

Using the soundtrack audio time line, the soundtrack producer can direct the recording session based upon the flow and mood intended in the script. This requires a sense of audio visualization, good communication, and direction to the voice-over talent. The voice-over talent often does not have any feel for the type of music or sound effects that will accompany his or her voice. While reading the script, they have no idea what visual material will be presented on the screen, and need direction to determine the pacing and style of reading.

The audio soundtrack producer also has to listen carefully to make sure all words are enunciated and pronounced correctly. For many soundtracks, this will require thorough up-front preparation to identify any words where the pronunciation may be unclear. Not preparing ahead of time may require a portion of the narration to be rerecorded at a later date due to incorrect pronunciation, requiring some type of additional fees to be paid to the voice-over talent, and the studio for rerecording short segments.

When laying out the soundtrack production for recording, it is often a good idea to carefully look at and review all aspects of an interactive production. In a well-designed program, the interaction could take the user many different routes within the program. Often audio will be used as a response to the program's viewer. It is often a good idea to record a number of stock library responses that can be digitized and be available on the system to prompt the user. Stock responses may include help or assist modes, or our example in the training application where we point out that, "Your answer is incorrect; the program will now take you back for a short review." A typical stock response for a point-of-sale application might include, "Sorry, that item is not available at this time." Again, it is more efficient to record the extra responses or pick-up lines you may need in the original recording session rather than having the voice-over talent return at a later date.

Mixing the audio track

Once the audio narration is recorded and edited to remove errors or bad takes, it can be mixed with music and sound effects. For many training and information applications, the audio track will probably be used by itself. Care should be taken during the mix to make sure the various audio elements are at the proper volume level. For example, music, when mixed with narration, should be brought in first at a higher volume level and then dropped when the narration starts. The effect catches the viewer's attention with the music, sets the mood, and transfers that attention to the words spoken by the voice-over talent.

Care should also be taken with sound effects. In our example of the opening to the Horizons module, the sound effects of the wind, seagulls, and the gasoline engine of the Wright Flyer were all mixed very subtly into the soundtrack. For some viewers, unless the sound effect is pointed out, it is very subliminal, but still adds to setting the tone and mood for what is appearing onscreen.

DIGITIZING AUDIO

For production of other media, the sound mix is the final stage of production of a sound track. With digital audio, a "completed" soundtrack must now be digitized. The ActionMedia 750 Production Tools include the software utilities to digitize the completed soundtrack into individual files for integration into your final application. The tools necessary for digitizing audio are illustrated in Fig. 5.5. They consist of your audio source which could be reel-to-reel, cassette, or DAT tape system. Or, a CD disk player could be directly connected to the ActionMedia 750 Capture board for digitizing music and sound effects. For recording and digitizing narration directly into the system, a preamplifier typically needs to be used between the microphone and the ActionMedia capture board to change the signal from microphone level to live level.

ActionMedia Production Tools include a software program called *VAudRec* for controlling the process of digitizing the audio soundtrack. VAudRec includes a simple straightforward character-based interface, as shown in Fig. 12.4.

The interface screen for VAudRec includes the options of setting the sampling rate, the name of the output file, and a readout of the frame count. The ActionMedia 750 Production Tools use the frame count as their principle navigation device. While at first this may appear to be difficult to understand, keep in mind that it is based on video working at 30 frames per second. Therefore, every 30 frames equals one second and every 1800 frames equals one minute of digital audio.

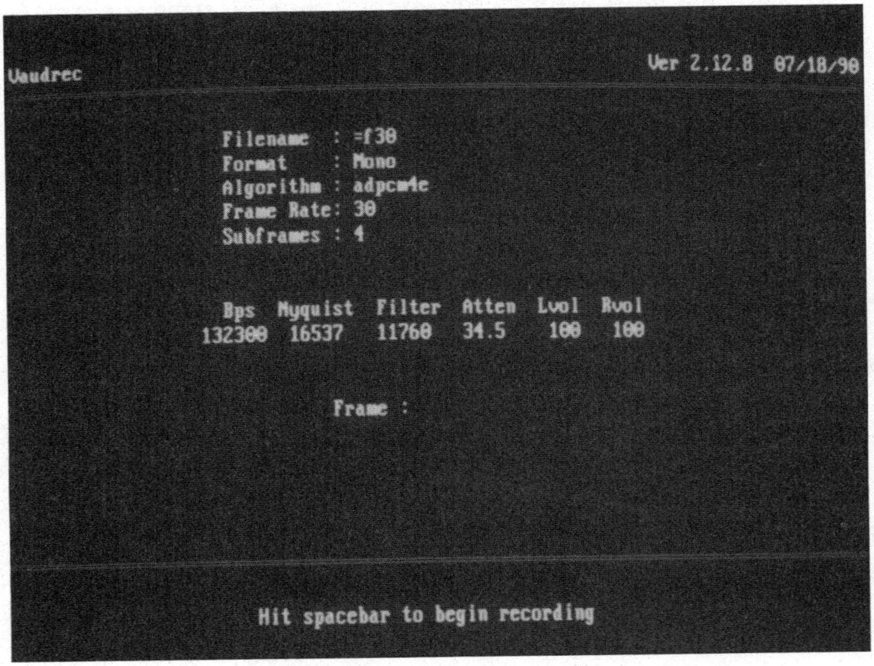

Figure 12.4 VAudRec Audio Capture Tool Interface.

Preparing for an audio digitizing session

We have found it best to do all of the audio digitizing for an application in one continuous session. With digital audio, the adjustment of levels is critical, especially with interactive digital audio where one audio file may stop and another may start immediately, based on the user's interaction. Small differences in the recording level will cause very apparent changes in the volume and the tone of the audio. The best recommendation is to either carefully document all of the settings for your audio equipment during a digitizing session, or attempt to digitize all of the audio segments at the same time.

Preparing your hard disk for audio digitizing

Next, your hard disk should be prepared for recording digital audio. It is recommended that you prep the hard disk to defragment the available disk space by using one of the commercially available defragment utilities contained in the Norton Utilities software from Symantec or PC Tools from Central Point Software.

If your working hard disk is not defragmented, you will often find that the audio during record will not retain a consistent 30 frames per second. The reason for this is very simple. During production, large data files for images and motion video are frequently added to the next available space on the hard drive. As the hard drive gets closer to filling up, there are many small spots of space. During an audio digitizing session, if the hard drive begins to record the digital data using many of these different sectors, the result is a delay while the hard drive jumps to those sectors in an attempt to record continuous data stream. When you defragment the hard drive, you compact and rerecord all of the previous data into continuous blocks. This makes files run faster, since the information is contiguous and the program will not have to "jump" all over the disk to find the data for each individual file.

Using VAudRec

Once you have the hard drive prepared and all of the settings correct, the recording process can be started by manually tapping the spacebar and starting your audio source at the same time. It is best to try to synchronize the starting point to allow the smallest amount of open space at the beginning of the audio file. This will keep your audio files smaller and more compact. It is also easier to synchronize the start of an audio file when programming or authoring the audio into your final application.

The interface for using the VAudRec software utility is fairly straightforward. On the main screen the sampling rate, filtering factor, Nyquist level, and stereo or mono can be selected. The source music can be played through the digitizer to hear the results of the effects of different sound levels and filtering adjustments. Once you are ready, the music can be queued up and the digitizing process started by pressing the spacebar on the keyboard. While digitizing, the frame counter will continue to count at the rate of 30 fps to give you an indication of how long the audio file will be in frames.

To edit or trim audio files, a software program called *VAudEd* is available in the ActionMedia 750 Production Tools. Using VAudEd, you can trim frames from either the front or head end, or the back or tail end of the digital audio file. The interface only gives you the ability to trim frames, and unfortunately does not have a graphic on-screen readout of the digital audio file. Retain a back-up file of your original digitized audio file in case an error is made during audio editing. (D/VISION, described in Chapter 11, can also be used for audio files.)

Playing back digital audio

To listen or play your digital audio file, a utility called *VAudPlay* is included with the ActionMedia Production Tools software. Listen to the playback carefully. Make sure that your manual synchronization at the beginning of the audio digitizing section caught all of the information at the beginning of your soundtrack. Also, watch that the record rate continued at 30 frames per second.

When first setting up a session for digital audio recording, you may also want to play back a previously digitized file, such as a com-

SAMPLING RATE

The equivalent of image resolution in audio is the *sampling rate.* DVI®'s programmable architecture extends to audio. The electronic component which controls the audio is called a *Digital Signal Processor,* or DSP for short.

To understand how digital audio is played back, let's first look at an overview of how audio is digitized.

When the audio enters the ActionMedia capture board, it is converted from an analog to a digital signal. The digital signal is analyzed every X thousands of times each second. X is the sampling rate. DVI® technology samples at four different levels: 4KHz, 8KHz, 11KHz, and 22KHz in stereo or mono.

Why the different sampling rates? Like image resolution, quality has its price in the resulting file size. A 30-second clip of audio digitized at an 11KHz sampling rate would result in a file one-quarter the size of the same audio clip digitized in stereo at 22KHz.

pressed video file, to judge the volume, quality, and tone to maintain a level of consistency with your application. Volume is especially important, as it can become a nuisance for users to have to reach over and constantly readjust the volume on their speakers because of mismatched audio in a application program.

By following your log of digital audio files, and stepping through a complete session of digitizing all of your audio, you will have the resources necessary for programming or authoring the final application. Maintaining strict quality-control standards and consistent levels will make sure the audio quality is maintained throughout the application. The results are enhancement to the application which initially may not be immediately apparent, but will add to the depth and dimension of your entire production.

Bringing Your Application Together with Programming or Authoring

The programming or authoring stage is where all of the elements of multimedia production are brought together to create the final application. This process of assembling all of the components could be likened to the final edit of a film or videotape, with the added dimension of control of the interactive process. Programming or authoring for multimedia also involves working with the rules defined for the assembly of objects. As we have seen in previous chapters, one of the keys to efficient multimedia design and development is keeping an eye out for ways to constantly economize the storage and display of the various multimedia elements. The programming or authoring task is no exception.

For example, to give the end user control over an application, a graphics bar with buttons already in place can be called from the hard disk or CD-ROM and "pasted" to the bottom of the screen. With good design and careful programming or authoring, it is possible to take the constantly used button bar and store it within the video memory of the ActionMedia board for quick display.

Another example of the tasks required in the authoring programming process is the assembly of sound elements with their associated graphics or stills. The programmer author queues the visual elements to their associated audio during the programming stage of the production process. Digital video segments with either the original audio or alternate audio soundtracks are also sequenced and assembled during this phase. In addition to assembling all of the elements, the programming or authoring process is where interactivity is implemented.

Selecting the right tools

It is best to select your authoring or programming strategy during the preproduction phase. In the DVI® application development and environment, there are four alternatives:

1. C Programming
2. Authoring
3. Scripting
4. Hypermedia

The four alternatives offer a variety of features, functionality, and expertise.

C PROGRAMMING

C programming for DVI® application development is one of the lowest levels, or more direct ways, to control the programming of a DVI® technology application. This low level offers the most functionality because it directly accesses the extensive capabilities the C Libraries provide in the ActionMedia 750 Software Library.

The C library consists of a series of software routines written to access many of the visual and audio capabilities of the ActionMedia boards. A sample of a segment of C Code is shown in Fig. 13.1.

C code allows the most flexibility in development, but it also requires a level of understanding of the C programming language and a series of new skills to understand how to call the necessary library routines to implement video and audio effects. There are numerous sources to learn more about C programming.

DVI® Technology's multitasker

Due to the word *multi* in multimedia, a part of the software architecture for developing ActionMedia applications requires the ability to start and control several operations at once. The DOS operating system for microcomputers was not designed with this in mind. Since

```
I16     menu()
{
I16     sel,  x,  y,   count;
GrPix   color;

/*   Load the font, if it has not been done yet  */
if  (text_setup == 0)  {
    x = GrFontOpen ("sans.112", NULL, &font_handle);
    if (x >= 0)  {
        x = GrFontLoad(font_handle);
        if (x >= 0)  {
            GrBmSet (pBM,    GrBmTextFont, font_handle);
            GrBmSet (pBMw,    GrBmTextFont, font_handle);
            }
        }
    text_setup = 1;
    }

/*  Draw the menu if that has not already been done    */
if  (menu_drawn == 0)  {
    /*  Draw a light blue background rectangle for the menu  */
    GrPixFromColor (pBM, GR_COLOR_RGB,  70, 85, 235, &color);
    GrBmSet (pBM,  GrBmDrawColor, color);
    GrBmSet (pBM,  GrBmDrawOutline, FALSE);
    GrRect (pBM, MENU_X, MENU_Y, MENU_WIDTH, MENU_HEIGHT);
    /*  Draw a 2-pixel wide yellow box around the menu    */
    GrBmSet (pBM,  GrBmDrawColor,  colors[1]); /* yellow */
    GrBmSet (pBM,  GrBmDrawOutline, TRUE);
GrRect (pBM, MENU_X, MENU_Y, MENU_WIDTH, MENU_HEIGHT);
GrRect (pBM, MENU_X + 1, MENU_Y + 1, MENU_WIDTH - 2,
       MENU_HEIGHT - 2);

/*  Set up the text to be transparent  */
GrBmSet (pBM, GrBmTextBg,  FALSE);

/*  Set up location to draw text from array: item[]  */
x = MENU_X + 10;
y = MENU_Y + 4;

count = 0;
/*  Draw items every 78 pixels,  until a nul item  */
while (  (count < 24) && (*item[count]  != (char) ' ' ) ) {
     if  (count == 12)  {
              y += 15;
              x = MENU_X + 10;
              }
```

Figure 13.1 Sample of Segment of C Code.

```
    /*  Draw the item number or letter in white  */
       GrBmSet (pBM, GrBmDrawColor, colors[0]);   /*  white  */
       GrText (pBM, x, y, item[count]);
       count++;
       /*  Draw the item name in black  */
       GrBmSet (pBM,  GrBmDrawColor, colors[7]);  /*  black  */
       GrText (pBM, x + 16, y, item[count]);
       x + MENU_PITCH;
       count++;
       }
     menu_drawn = 1;
     }

  /*  Now wait for a keystroke  */
  sel = getch ();
  /*  return 0 to quit if ESC or Q or q is hit  */
  if  ( (sel == 27) | (sel == 81) | (sel == 113) ) return (0);
  return(sel);
  }
```

Figure 13.1 *(cont.)* Sample of Segment of C Code.

DVI® products run under DOS, the ActionMedia software environment includes a real-time executive program called RTX for short. The RTX code allows a number of activities to be started and tracked simultaneously, including the display of visual information, and the start and stop of a video soundtrack. Often, an application design will also require the computer software to be able to check or poll a mouse device or keyboard input device to determine if the user has a response or wants to interrupt the process. The skills for programming with RTX are documented in the ActionMedia 750 Software Library documentation.

The skills for programming multimedia in the C language and using RTX are developed through a combination of training and experience. It is recommended that C programmers planning on writing code for ActionMedia software take the C level programming courses offered by Intel to understand this process and to maximize the capabilities of the ActionMedia programming environment.

AUTHORING

C programming, while offering the maximum flexibility, can also be more costly due to the special skills required of a C programmer and the time it will take to write code required for an application. Authoring products have been developed to simplify the multimedia

ACCESSING THE i750 VIDEO PROCESSOR THROUGH MICROCODE

The i750 video processor on the ActionMedia boards is separated into two separate microprocessor chips—the *Pixel Processor* (PB) and the *Display Processor* (DB). These processors were designed with the specific job of compressing, decompressing, and displaying video information. Like other microprocessors, the i750 PB/DB require input, output, some memory for storage, and control circuitry. A key part of the design of the i750 PB is its onboard RAM. This RAM, built into the processor, allows instructions or small computer programs to be loaded directly onto the component to perform video processes. These instructions are called *microcode.*

Good examples of the use of microcode on the i750 PB are video effects. To understand how microcode works, we will use the dissolve effect from one 9-bit still to another as an example.

Computers are known for their ability to tirelessly perform repetitive processes over and over again. A video dissolve is an example of repetitive activity being performed by the i750 video processor. A dissolve occurs between two images when one is being displayed, and the other is loaded into the VRAM available on the ActionMedia 750 Delivery board.

The action of the dissolve at a pixel level is the mixing and replacement of the color values on a pixel-by-pixel basis. For example, if the first picture has a red background, and the second picture has a blue background, the dissolve microcode routines calculate the range of color values between the red and blue colors and adjust for the display of each color over a specified time span. This process is repeated sequentially across 512 pixels per line, and a total of 480 times for each line of video displayed on the screen. With this example, you can see the extent of the calculations required and repeated every time the screen is refreshed (over 30 times each second), and the new image displayed. The i750 video processor manages all of this computation from its memory.

Intel has developed a library of microcode routines for most common video effects and transitions. With a fair amount of preparation and understanding, it is possible to write a new routine for a video effect that does not currently exist. One example of this is a new microcode routine developed by Time Arts for their popular Lumena for DVI® technology paint program. Time Arts developed a routine for an airbrush effect. Real-time airbrushing on-screen requires a lot of calculations to create a smooth, but random, display of color pixels onto a screen. This controlled, but random, effect requires a tremendous number of calculations, as the display must react quickly enough to be used in a stroking motion, the same way an artist would use it. Previous paint programs that have relied on a personal computer's microprocessor have typically lagged to the point where it was difficult to offer an electronic graphic artist an effective airbrush tool. Using the i750 video processor's fast calculation capabilities, the airbrush effect is very similar to the conventional paint/spray painting technique.

Writing microcode typically requires the skills of the specialist. It is not usually done by an applications programmer, nor is it needed for application development. We discuss it here as a matter of interest, and to illustrate the enormous flexibility available to developers working with the i750 video processor architecture.

programming process by offering access to the most-often-used capability of the ActionMedia environment.

Authology:MultiMedia

The first example of such an authoring tool is Authology: MultiMedia from CEIT Systems. Authology:MultiMedia uses two primary metaphors for developing applications. The panel editor is the interface for bringing multimedia elements together into a series of frames or panels. For example, using the panel editor, instructions can be given to the program to load a background image file from CD-ROM or hard drive storage. The bit maps for individual elements such as buttons can also be called up separately and displayed on the screen. The panel editor allows the authoring programmer to move bit-map objects around on the screen to their appropriate positions. Text can be displayed by the program during playback. The panel editor also allows primitive graphic objects such as circles, squares, lines, and polygons to be defined and displayed to the screen. A panel, as shown in Fig. 13.2, may consist of a series of stills leading up to a clip of motion video which can be played either full-screen or in a portion of the screen, all under control of the authoring environment.

As shown in Fig. 13.3, the procedure editor allows the various elements of the application program to be linked together through a

Figure 13.2 Authology:MultiMedia Panel Editor for Creating Graphic Screens.

"point-and-shoot" interface. Here, the developer can select from pull-down lists of the next available commands to initiate an action. When that action must, for example, display an image file, a pop-up menu allows you to select from those image files stored together for this application. The result is an organized methodology for developing an application with little need for on-screen keyboarding. The benefit, well understood by those who have typed in traditional program code, is a dramatic reduction of program errors or bugs due to typos and syntax errors. Once the audio visual panels are developed, they are linked together in a procedure mode.

Authology:MultiMedia offers access to most DVI® capabilities. For many of the typical training and point-of-sale applications, Authology:MultiMedia will meet all of an application developer's needs.

Occasionally, an application developer will want to access other special capabilities; for example, the analysis of the input from a control panel for simulation applications. CEIT Systems has developed a means for accessing outside software programs, or *modules,* written in C that allow a developer to put together the specific codes for updating information from an external control panel and interface.

To distribute Authology:MultiMedia applications, a run-time version of the program is available. The run-time version is the software player that is optimized to include only the code necessary for playing back the commands compiled by the Authology:MultiMedia authoring

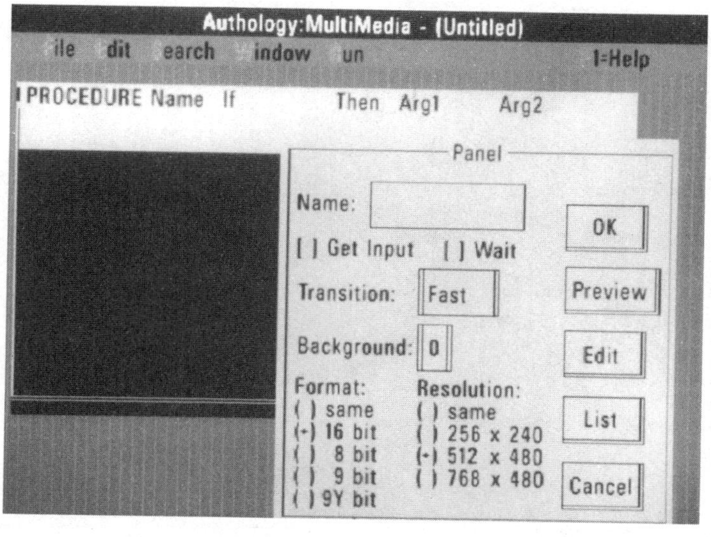

Authology:MultiMedia - (Untitled)
ile dit earch indow un 1=Help
PROCEDURE Name If Then Arg1 Arg2
Panel
Name:
[ ] Get Input [ ] Wait
Transition: Fast
Background: 0
Format:
( ) same
(•) 16 bit
( ) 8 bit
( ) 9 bit
( ) 9Y bit
Resolution:
( ) same
( ) 256 x 240
(•) 512 x 480
( ) 768 x 480
OK
Preview
Edit
List
Cancel

Figure 13.3 Panel and Procedure Display for Ceit Systems' Authology:MultiMedia.

program. The run-time version carries a nominal per-unit charge, with quantity and corporate distribution licenses available.

MEDIAscript

MEDIAscript offers a different approach to application development by simplifying the multimedia application programming process through a scripting language. This scripting language can be compared in concept to the same approach used in the highly popular Hypercard program for the Macintosh. It allows a developer to quickly prepare many of the application code routines for multimedia development with a few, brief commands. The code for a MEDIAscript application is shown in Fig. 13.4.

```
*  MEDIAscript™ Sample Code
*  Displays motion video file in Window which moves around
   the screen

cls ltblue

*  load the "Actionmedia" image and get its background color
i actmed5
s dest=1:grab 1 1:s bkgnd=gcolor,dest=o
*  copy it transparently to the screen
ict im 29 100

*  load the arrays for the animations
i array,colt0314/0,480

*  Now do four colt animations off-screen
s dest=1
task0 seq im 128  120  128    0  128  120  0  2
task2 seq im 128  120    0  120  128  120  0  2
task3 seq im 128  120  384  240  128  120  0  2
task4 seq im 128  120  256  360  128  120  0  2

*  Now animate the moving window to the screen
s dest=0
d 2  1  1  -2  -2:r   60  278  388  124  yellow
task1 seq 0  0  511  480  1  120   62  280  384  120  0  2/r

*  wait for a keystroke or click to end it
mouse
term
```

Figure 13.4 Sample MEDIAscript™ Code.

When compared to writing DVI® applications in the C language, you can see the simplified approach offered by MEDIAscript. Much of the computer environment housekeeping required for C is taken care of for you by the MEDIAscript programming environment. The intricacies of working with RTX are all managed transparently in the background. The developer can concentrate on accessing those capabilities of MEDIAscript to create the desired audio and visual effects.

MEDIAscript was developed specifically for DVI® technology by Arch Luther, one of the original members of the development team at the David Sarnoff Research Center. MEDIAscript offers a great deal of flexibility when developing DVI® multimedia applications.

Included with MEDIAscript are a number of easy-to-use tools that simplify the application development environment, such as a frame-counter, or a pixel editor that allows you to quickly go in and touch up (on a pixel-by-pixel basis) DVI® images.

There are a number of tools and utilities that are being developed by outside third parties in the MEDIAscript environment for DVI® multimedia developers. These include video editors and audio editors with easy-to-use "point-and-shoot" interfaces that allow you to access the capabilities of DVI® technology without C code or command-line style utilities and applications.

MEDIAscript also offers a run-time module optimized for playing MEDIAscript applications. This run-time module can be distributed with the application, and is available at a nominal cost in single units, quantity licenses, and for corporate site licenses.

Hyperties

Many applications are text-oriented and follow the metaphor of a book. Here, multimedia is used to support the written word. To support applications that are heavily text-oriented, Cognetics Corporation has developed a program called *Hyperties* for DVI® technology. The Hyperties program is interesting in that it allows the application developer to link words within text to multimedia objects and to each other. The result is an enriched electronic document providing multiple levels and types of information resources available to the reader.

The Hyperties program was originally developed as a text-only-based program to create linkages within a document or series of documents for a CD-ROM. For example, if a repair manual showing how to fix a car mentioned carburetor, and the carburetor was highlighted, selecting the word *carburetor* with a mouse would provide a link to "jump" directly to the portion of the document that provides more information on carburetors. This cross-linking power is the heart of

Hyperties. The authoring environment includes tools for scanning and building links to word associations.

Let us take, for example, a training document on DVI® technology itself describing the ActionMedia 750 Display board. Within the text of product specifications are a number of terms and product names that are cross-referenced. In our example, Fig. 13.5, we see two examples of highlighted text on-screen. The highlight indicates that more information is available; in this instance, a multimedia element is available for display.

The Hyperties approach applies a completely new concept to the development of multimedia information. The book metaphor is one we know well. Text information can easily be converted into electronic form, and with the addition of products like Hyperties, links to multimedia elements can be easily accomplished. The result is an enriched environment where additional information is available for those who need or desire it.

Cognetics, the publisher of Hyperties, offers a run-time version for distribution of application. The run-time version is available at a nominal cost with site licenses available.

SELECTING THE RIGHT PROGRAMMING OR AUTHORING ENVIRONMENT FOR AN APPLICATION

As you can see, all four of these different choices for programming and authoring offer specific benefits and advantages, depending on the application. Selecting the right approach starts back in the preproduction process, and is based on the number of qualifying factors.

The choice between C and authoring

C programming offers the maximum flexibility. This flexibility is offset by the programming expertise and time required to program C language applications and the additional knowledge necessary to understand the multimedia capabilities available to the C programmer through the ActionMedia 750 Software Libraries.

The application developer also needs to look at how the program will interact with other input or output. Products such as Authology:MultiMedia, MEDIAscript, and Hyperties are all designed for the majority of applications, but do not take into account the special capabilities of DVI® technology available for custom applications. This point alone may be enough to force the decision toward the C programming environment.

For most applications, Authology:MultiMedia or MEDIAscript are well-suited to the task. MEDIAscript offers more flexibility and

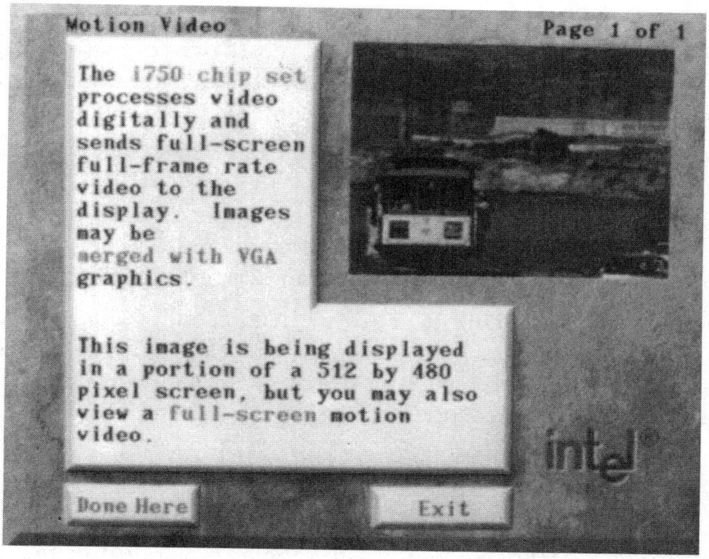

Motion Video Page 1 of 1

The i750 chip set processes video digitally and sends full-screen full-frame rate video to the display. Images may be merged with VGA graphics.

This image is being displayed in a portion of a 512 by 480 pixel screen, but you may also view a full-screen motion video.

intel

Done Here Exit

Figure 13.5 Hyperties Application with Highlighted Terms Indicating the Availability of More Information.

access to more capabilities than Authology, with the tradeoff being the difference of a "point-and-shoot" interface and scripted commands that must be typed in from a keyboard.

Which programming environment to choose

The decision of which authoring package or programming environment to use many times falls to personal preferences. Some application developers feel a "point-and-shoot" interface gets in the way and slows down application development, while others would like the structured approach of a program like Authology:MultiMedia. This question is a difficult one, and can be influenced by a number of factors:

- Experience of the team members is a consideration. If you have an author or programmer who has used a particular environment before, or similar environments, this may constitute a preference.
- Efficiency is also a consideration. Again, if someone on the team has experience with one of the authoring programs or software that is similar, he or she may be able to complete programming more quickly and with less learning time.

- Capability can sometimes be a factor. MEDIAscript offers some tools that allow programmers to allocate video memory. Authology:MultiMedia manages all the memory for the author, and does not allow the author access to memory management. The C language offers complete flexibility in this area.
- Speed of operations is most efficient when writing C, but this is only manifested in instances where image manipulation is used extensively, like in unwrapping fisheye images, or panning and scrolling through large images or complex animations.
- Economical factors can also apply. If the intended audience already has an installed base of the run-time modules of either of these programs, or if a favorable run-time license can be negotiated, you may consider one over the other.

Text-based applications

If your application consists of a lot of text, the decision is much more straightforward. Hyperties offers the tools to easily link text to text, and text to multimedia elements.

Other multimedia development tools

If these application development environments are not suited to your needs, others are on the way. New authoring programs and environments are being developed to take full advantage of the graphics display capabilities of DVI® technology. In fact, depending on the needs of your application, a new industry of third-party software developers is emerging with tools for DVI® multimedia application development. This includes video editors from TouchVision and Montage, and sound editing tools from Network Technologies, the publisher of MEDIAscript.

Other operating environments coming

In the near future, other operating environments will be supported such as Windows, OS/2, the Macintosh, and Unix. Tools are already under development to take special advantage of the highly graphical interface these environments offer.

THE PROCESS OF AUTHORING OR PROGRAMMING

Authoring and programming immediately conjure up the image of linking all the multimedia elements together into a running program. In reality, this is only about 25–35 percent of the task. With a well-organized flowchart and a good understanding of how to use the authoring tools or C language, an application can be programmed to run in a relatively short period of time. Typically, a programmer will write a skeleton program as a first step. This program, when it is completed, is a sort of prototype of the actual program, and can be used with substitute elements to try out effects or interactions. In the beginning, you should not expect to actually "see" much as the skeleton program is being built. This step is like preparing the field for the big game, or digging the hole for a skyscraper. It is the foundation for something big, and requires a lot of work, but doesn't look like much to someone who isn't an expert, or who hasn't done it before.

Most of an application programmer's time, however, is not spent on this task. Rather, it is spent on adjusting, positioning, and timing the various multimedia elements to come together with the look and feel originally planned in the initial design. For example, playing a sound effect at exactly the right point in time is a key part of the programming and authoring effort.

A good design will indicate timing points, but in the actual development environment the programmer often finds that trade-offs and adjustments have to be made to reach the desired effect. One approach to the development process at this stage is to have the programmer or author code the application straight through without paying particular attention to timing. This stage of work, similar to the rough cut in film or video production, can be reviewed with the design team. Often, it will be evident when effects do not work, or where transitions may need to be changed, or where a portion of a program is missing or does not work well. After the rough cut, the programmer can focus efforts on these areas to fine tune the final program.

Our example of stills with audio build can be used to illustrate this concept (shown in the color figures). The first pass of programming showed that the transitions with the seagulls on the screen in one group did not create the desired effect with the selected music. It was more effective to bring the seagulls up one at a time from right to left as the sound effect of the seagulls increased in volume.

When this was discovered, two techniques were discussed to correct the problem. The first was to transition from several full-screen images, each one containing one more seagull. This was considered inefficient, since it would require five full screens to be stored. The speed of the display would be too difficult to maintain with five full transitions in the desired time frame. The more efficient method, and the one used, was to display a base background image and transition only between the individual cut-out bit maps of each individual seagull. Even with this seemingly small change, the performance and speed of the transition was affected, and the timing of the audio and still-image display had to be adjusted. Many times, with more elaborate interactions, a transition or change in display must be adjusted, played back, and readjusted several times to achieve perfection.

Many programmers and authors learn the "tricks of the trade" with experience with a particular language or authoring system. They will often build a library of tools, or code templates, that will allow them to build a rough cut quickly. With good planning, these templates can be built so that timings, changes in elements, and interactions can be changed relatively easily. However, there is no substitute for giving a programmer a complete flowchart, storyboard, and allowing lengthy discussion of how the program and interface should work.

Preparation of media elements

Good preparation of media elements can reduce the time a programmer or author needs to spend in the final stages of production. A production assistant or other team members can prepare and edit all the media elements so that the programmer is receiving final versions of motion video files, audio files, and graphics.

For example, one technique available to programmers is to store all the menu buttons or interface graphics as an image in off-screen memory. As these graphics are needed, the software can display them in the proper place on the screen. If the programmer is given a completed graphic or storyboard showing where buttons or pull-down menus should appear on the screen, the programmer can spend time efficiently designing how the menu can work, rather than designing where the buttons should go, and inevitably change the interface each time a design review cycle is held.

THE PROGRAMMER AS TEAM MEMBER

If you plan to add to a program in the future, the programmer should know where, and what type of interactions you will expect to add. If you plan to distribute your application on CD-ROM, hard disk, or other media, the programmer needs to know this for planning purposes. In short, the programmer should be a well-informed team member. As we have illustrated many times in this book, almost all the decisions made about production have a technical consideration.

An experienced programmer or author can contribute to the development team in almost every phase, from design to production and distribution. There is probably not anyone on the team who will be able to advise you better about what features are available, and what creative alternatives you have as a DVI® producer.

Section

3

Getting Your Application Ready for Distribution

Chapter 14

Testing Your Application

Application testing is often a step that is forgotten or shortened in the application-development and production process. No one likes to admit that their finished application may actually have bugs or errors. Experience shows that unfortunately there usually are a few bugs lurking within a program.

DEVELOP AN APPLICATION TEST PLAN

Each application, due to its different requirements, should have an *applications test plan.* This plan should be monitored and documented to make sure that each area of the program is properly tested. Problems during the application-development process often do happen. They can be very costly to repair after a program is released. The following should be part of an application-development plan:

Functional tests of all steps within the program.

Test the installation procedures.

Random tests.

Internal alpha test.

External beta test.

Follow up the final release.

Let us take a look at each of these areas to establish a testing procedure.

FUNCTIONAL TESTS

The first part of establishing testing procedures is the assignment of this task. It is recommended that this testing is not done by the same programmers who write the initial code. During the application-programming process, testers should exercise various subroutines, as well as the entire application. The functional tests are the first step in the verification process to determine that all of the elements of the program work as planned.

Someone should be assigned to actually implement the tests, and establish and document the testing environment. Pay particular attention to the configuration. Check the configuration used by the programmer during application development for tests. Review this configuration with other common programs to determine whether it is consistent with the configurations recommended for running DVI® applications. (Appendix C contains a list of common configurations.)

During the functional tests, it is best to use the flow chart of the product as the guideline for testing. Using the flow chart for documentation, each branch of the program flow can be tested along with its functionality. The objective is to functionally test each and every capability of the program to make sure it works. Mark off each leg as it is tested to document that it has been tested and works.

Bug report forms

A form similar to the example shown in Fig. 14.1 should be filled out each time an error or anomaly is found. It is important to know the nature of the error, and any symptoms that led up to it. This information gives the programmer clues as to where the problem might be when attempting to resolve it.

Bug reports should be cataloged and noted together. In an ideal situation, a database should be developed to track the bug and potentially indicate any commonality that may help in tracking down a persistent and nagging software problem.

Bug reports should remain "open" until the problem is resolved. It is also important that bugs be prioritized. The following level recommends definitions for prioritizing bugs.

Level 1—crashes or hangs program—must be fixed before release

Level 2—functionality does not properly work—should be fixed prior to product release

DVI Application Development Bug Report Form

Bug ID		
Tester Name	**Phone**	**Date**
Short Name of Bug	**Module/Function/Section**	**Date Resolved/Who**
Bug Type	**Priority**	**Status**
Bug Description		
How Produced		
Suggested Resolution		
Potentially Related Issues		

Figure 14.1 Sample Bug Report Form.

Level 3—functionality that needs to be implemented before release

Level 4—feature which should be added to this release or the next release

Using a bug priority scheme will allow the programmer to manage time better, and to focus on the most important bugs first.

The ripple behind a bug

The process of tracking down and fixing bugs can be very frustrating. The biggest concern is that a bug fixed in one location of the program will cause something else to break somewhere else in the program. An experienced tester is aware of this problem, and will look for symptoms of where a bug can surface again.

Closing bug reports

Only when the tester has been satisfied that the bug and its associated symptoms have been fixed or resolved, should it be closed on the Bug Report form. The Bug Report form is also important, as it allows information from the testing process to be distributed among a group of people working on a project. This communication can be important, as bugs can be caused by some other problem, potentially in the operating system, or even in the hardware. Sharing this information across a work group allows the team to share their common experience in the detective effort sometimes required to resolve persistent problems.

INSTALLATION PROCEDURES

Once you are satisfied that the basic operating code works, it is a good idea to develop installation procedures for further testing of the project. Because multimedia utilizes a number of peripheral devices, it is important that your application software program be delivered with the appropriate drivers or test routines to determine that the drivers are available in the environment.

A number of very well-designed installation programs are available from software shops who sell primarily to software programmers. These programs have the functions built in to check the computer's environment to make sure that it matches what you need for your application.

For the DVI® environment, you want to make sure that all of the necessary load files are contained in the "config.sys" and "autoexec.bat." Good industry standards discourage automatically copying over the end user's existing configuration files, as this will often delete valuable information, and the application developer is not always aware of what additional software options each individual user has installed on his or her system. Well-done installation programs will add the new drivers to the appropriate point in the configuration and batch files. They will note possible conflicts and even recommend a resolution.

"RAM cram"

Available memory is of particular concern when running DVI® applications within the DOS environment. Often, users will have a number of driver programs or terminate-and-stay resident programs on their systems for the various utilities they run every day. Your installation program should detect the amount of memory available, to determine if there will be a conflict later on when you attempt to run the application.

If your program is to be run from a CD-ROM, you will want to make sure that the CD-ROM is installed, and determine which disk drive it has been assigned to. In addition, you may elect to offer the user the ability to copy many of the commonly used files for a multimedia application from the CD-ROM to the user's hard disk. This will increase system performance and response time.

We recommend that you transfer run-time programs and the primary instruction files for the applications over to the hard disk when a program is distributed on a CD-ROM. This will increase the speed and performance of the multimedia program. If additional room exists, give the end user the choice of dedicating additional hard-disk file space for the application. Performance can also be improved by copying over primary menu files, backgrounds, and icon or button files. All of these will increase the performance of the application, since they are the most frequently called files in most application environments.

It is best to add the installation procedures now to begin to establish a standard environment for the next stage of testing.

RANDOM TESTING

The next testing operation is to begin to randomly test the application similar to the way in which an end user would use the program. It is recommended that this testing start from the beginning with installation to develop and maintain an environment as close as possible to what the program might encounter when distributed in the field. Random testing begins to apply real-world situations in the test program. Your testers will be watching for memory conflicts, or situa-

SYSTEM CONFIGURATIONS

The DVI® development environment adds a number of interesting new capabilities which must be controlled within DOS. Naturally, these new capabilities require additional system RAM memory. At the same time, each application developer faces the situation where they need the maximum amount of memory for their application. The end result has coined an industry term called *RAM CRAM*.

While RAM conflict issues will eventually go away under advanced operating systems, such as Microsoft Windows in enhanced mode and OS/2, it is a problem we need to face today. For example, Appendix C of this book includes three of the optimal configuration files for the application development tools: Lumena for DVI® technology, Authology: MultiMedia, and MEDIAscript. In addition, a fourth configuration is the application playback environment recommended by Intel. You will note that they are different, and not all programs will work in the same environment. This will require some testing and configuration work to determine the optimal environment for your application.

Avoid straying too far from the recommended configurations. These configurations have been tested under a variety of circumstances and in a number of installations for best results.

tions where sections of your applications code may have errors in the memory or display. Random testing will also begin to show areas where there may be problems with the basic design of the program. For example, random testing may show that functionality may not work exactly as expected. A program may be less responsive than planned, or may introduce "wait" time where it is not desirable. While these areas may end up as noted as Level 3 or Level 4 bugs, they are valuable input to the production process.

During the random testing, often the functions of the flow chart are followed, but in a more random pattern.

ALPHA TESTING

During the alpha testing, the program is actually placed into the hands of the internal users. The testing process is now moving closer to the type of user the program will encounter in the field. Alpha testing is another level of random testing where the program is in the hands of the typical "friendly" user. Again, particular attention is given to installation and configuration problems, and functional improvements that were not considered during the design process.

It is important that the application developer maintain control of the testing process. Many times, alpha and beta testers are volunteers, willing to work with the program in return for an opportunity for a first-hand look at a new software application. Because of the "volunteer" nature of the testers, it is important to provide them with a deadline date for feedback to keep the project on its intended production schedule. Sometimes you will need to find some way to provide an incentive to alpha and beta testers to make sure they dedicate the time to testing the application and providing feedback and bug reports on a timely basis.

BETA TESTING

The beta test stage is important, as it is the last step before shipment to actual end users. The traditional definition of beta testing is the point at which the product is shipped to actual customers who agree to test prereleased software. For this test to have full value, it is important that all identified Level 1 and potentially Level 2 bugs have been resolved during the previous testing stages. Because the program is going out to actual end users, every effort should be made, including in some cases the final packaging, to test all aspects of your application. Where known bugs are in the process of being resolved, indicate work-arounds or identify the bugs, so as not to waste a beta tester's time. Plan your beta test time for scheduled reporting points where the end users will have to be contacted to determine their initial impressions to the program, and to spot any bugs or problems they may have had.

Watch for hardware-related problems. During the beta test, you probably have lost the ability to control the end user's hardware environment. The end user may have special software located in her or his environment, or an unusual hardware configuration which could conflict with your application software. This is especially important, as the real world consists of a proliferation of clone computers, various peripherals, and many unusual software configurations.

ONCE YOU HAVE COMPLETED TESTING— THE REAL BETA BEGINS

Once you have passed the testing phase, the real beta test begins when the product is shipped to a large base of users. Just as in the beta test, your application will be faced with a variety of different hardware and software configurations people install on their personal computers. Also, you can just about plan that your program will be used in ways you had not even considered! Button combinations will be selected in ways that were never thought of, or a grouping of selections may take an end user in a direction you may never have intended him or her to go.

Good planning includes a budget for the human and financial

VERSION CONTROL

During programming, and especially during testing, version control becomes a critical issue. Nothing can be worse than testing a mixed group of versions of the software program. This clarifies input and allows the entire production team to communicate efficiently and effectively. Most software development programs begin by numbering the revisions or versions of their software using the first version of the software to be officially shipped to customers as Version 1.0. During the application development phase, the traditional numbering system starts at Version 0.10 and begins to build from there. Often versions are noted also by their release date or a build number.

It is not unusual for a program to have many, many releases during the development phase. Do not be surprised if the versions released are in the double digits.

It can also be very helpful to keep track of who has copies of the program and their current version. As bugs and problems are resolved, this provides a distribution list to keep everyone current with the development effort. This will tremendously aid the communication flow problems back into the production process. The value of bug reports is greatly diminished if the information is based on a version of the program three or four revisions old.

resources to update the original program shortly after the application is first shipped to end users. Protecting yourself for this contingency is a good precaution in case of an unknown catastrophic Level 1 bug requiring a very fast turn-around and release of the new program, or in the more typical case, a later release of the program to fix several annoying points or Level 3 and Level 4 bugs or design changes.

Because of the number of options and selections available by virtue of adding the additional elements of multimedia, you can assume that several months after release of the application, there will be a new input on how you could have done it better. Planning for an update release as part of the production process from the beginning will save you from a lot of heartache, and give everyone a realistic expectation of the scope of the project.

Chapter

15

Distributing Your Multimedia Application

Multimedia offers an additional challenge to application producers and developers when it comes to distributing applications. When compared to yesterday's character-based PC computer systems, the information-rich multimedia files, sometimes as big as hundreds of megabytes, require new ways of thinking about how to distribute applications.

In this chapter, we are going to provide information about several different methods and media for distributing multimedia applications. These include methods that can be used to distribute and playback applications:

1. CD-ROM
2. WORM or erasable optical disk systems
3. Local Area Network (LAN)

The methods that are best to use to distribute applications that will be played back from hard disk or other source:

1. Archive streamer tape
2. DAT tape
3. Bernoulli removable cartridge systems

4. Data compressed onto floppy disks

As is typical with most of the production processes for multimedia, each of the choices carries different advantages or disadvantages that must be weighed carefully. In most cases, the choice will come down to economics and the quantity to be distributed.

CD-ROM

The CD-ROM offers a number of advantages for storing and distributing multimedia applications. First and foremost is its relatively low cost for both the CD-ROM players and the disks themselves.

The CD-ROM drive systems have been rapidly dropping in price, thanks to the popularity and manufacturing similarities with the audio CD player. CD-ROM players are available for less than $500 and in the not-too-distant future should be available for almost half that cost.

The replication of CD-ROM disks has also dropped thanks to the manufacturing capacity available for standard audio CD replication. In quantity, disks can be produced for under $2.00. Mastering and set-up charges will add a flat up-front $1500–$2000. For distribution to 50, 100, or several thousand sites, it is difficult to beat the economics offered by the CD-ROM. The 660 megabytes of data storage appears to be just about the right size for most applications.

The trade-off—data rate

As we have seen throughout the multimedia production process, almost everything has a trade-off. While the CD-ROM is economical and offers the capacity of storing large amounts of data, the data rate and access time are relatively slow. The access time, or the time required to move the laser to the specified sector on the CD-ROM and read data, is slower than almost all hard disks and floppy drives. This is an important consideration during all phases of the application design. The testing process should also watch for gaps when reading files from CD-ROM. Many times these gaps can be covered by preloading or buffering an image or audio file before playback. Good design many times will overcome the performance issues of using CD-ROMs.

The High Sierra or ISO 9660 format

An international standard governs the formatting of data for replication on CD-ROMs. This format is called the *High Sierra format* or *ISO 9660* which is recognized by the International Standards Organization. Most CD-ROM drives are shipped with Microsoft CD-

ROM extensions (MSCDEX). Using the MSCDEX drivers, a CD-ROM looks like any other data-storage device in the DOS environment, except you cannot write data to it. You can change to the drive letter the CD-ROM is assigned to, and run a directory command. Files can be copied directly from the CD-ROM to the hard drive or to floppy.

Formatting to the ISO 9660 or High Sierra format requires special software. Companies such as Meridian Data Systems sell software for formatting data files to the ISO 9660 standard. Another alternative is to have a CD-ROM pressing facility format the data for you.

Preparing data for mastering a CD-ROM

A number of steps should be completed to prepare your data for CD-ROM mastering. First, it is best to organize your data into a series of subdirectories. This keeps all of the associated files together and should improve the relative access time. Next, eliminate those files not necessary for the application. Many times during application development, working or temp files will be created to store data such as temporary image or audio files. Do not copy any of those files as they will no longer be needed when the data is read from the CD-ROM.

The ActionMedia 750 Production Tools contain a software utility called *VLayOut*. VLayOut is used to pad files with additional null data. This "pad" data consists of null characters to fill out the data file so that it can be read at the CD-ROM's constant data delivery rate of 150,000 bytes per second and a consistent 30 frames per second.

Transferring the application's data to transportable media

Once all of your files have been prepared for mastering, they can be copied to:

An archive tape backup system

Archive DAT system

One of several approved write-once or erasable optical drive systems.

Not all systems store enough data to match the capacity of a CD-ROM. Several tapes or cartridges may need to be properly labeled and prepared for sending to the CD-ROM disk-pressing facility.

Packaging for CD-ROMs

One often-overlooked, last-minute detail is the packaging and labeling for the CD-ROM. Most CD-ROMs are delivered with a clear plastic, covered "jewel" case. The "jewel" case exposes the top, non-reading portion of the CD-ROM. Most disk-pressing facilities include in their prices the ability to silkscreen two colors of artwork onto the disk for identification and display. When placing your order for disk replication, the artwork for the disk must be submitted at the same time. Many application developers include logos and other artwork that must be prepared conventionally for silkscreening onto the disks.

CD-ROM mastering

Most CD-ROM pressing facilities allow for several different mastering schedules ranging from 10 days to same-day turnaround. Carefully determine your needs, as the prices vary dramatically in accordance with the speed of turnaround service.

Proof or check disks

A limited number of proof or check disks are pressed for testing your application after mastering onto the CD-ROM. Care should be taken for a complete functional and random test with particular attention paid to the load time for still-image and audio files. If there are long gaps as a result of the slow access speed of the CD-ROM, the programmer may elect to change the program to preload some data files. Part of the price for the mastering process allows for the cost of one check disk and one set of changes. Additional rounds of check disks can be ordered at an additional cost.

WRITEABLE CD-ROM RECORDERS

One recent development is the availability of CD-ROM ISO 9660 format compatible writeable recorders. These systems are currently available in the $25,000 to $30,000 range and allow you to create your own CD-ROMs. CD-ROMs created with this process are not molded or manufactured like the conventional CD-ROM. The process requires media costing about $75 a disk that allows the laser within the CD-ROM recorder to burn digital pits or holes directly into the CD-ROM's internal coating.

The CD-ROM recorder is a good device for preparing proof or test disks. Because this capability can be located within your production shop, test disks can be created quickly and tested. The test or proof disk can be played back from any CD-ROM player with the same performance characteristics as a conventionally produced CD-ROM. For some applications where there is very limited distribution of large amounts of data, the CD-ROM recorder, even with its higher-priced media, may be the ideal solution. There are also some organizations with security regulations that cannot send data off the premises for mastering. Again, the more expensive CD-ROM recorder may be the answer.

WORM OR ERASABLE DRIVES

WORM drive systems have continued to drop in price over the last few years, making them practical for storage and distribution of large amounts of data. Many of the CD-ROM pressing plants will accept a WORM cartridge for CD-ROM replication. WORM drives holding from 200 to 500 megabytes of data are available for less than $2,000 with media cost of between $100 and $125. Like the recordable CD-ROMs for limited distribution or applications where security requirements specify on-site replication of disks, WORM drives may be an alternative.

Erasable optical data storage drives have now started to become available on the market. The most popular utilize the magneto-optical method for recording and erasing data on removable cartridge optical disks. Drives for storage of 300 megabytes of data per side on data cartridges costs about $3,000 to $5,000. The data cartridges are in the $125 to $150 range.

Many application developers are finding magneto-optical systems the ideal format during the production process. Their performance is similar to a hard drive, but with the advantages of removable, erasable media where an entire application's working files could be stored on one or two cartridges.

LOCAL AREA NETWORKS

Probably one of the most exciting distribution methods is the use of a large database of multimedia information delivered to workstations through a local area or wide area network. IBM has publicly demonstrated a local area network running six or more multimedia workstations. This is an exciting development, as it holds out the true promise of "information at our fingertips" by allowing information to be potentially accessed from multimedia databases worldwide.

ProtoComm, an early DVI® software developer, has a series of software development tools available for developing multimedia applications to run from a Novell local area network. ProtoComm's VideoComm product offers special features which provide the capability to read DVI® AVSS files from a central file server at a constant data rate necessary for the display of digital full-motion video.

ARCHIVE STREAMER TAPE BACKUP SYSTEM

The Archive streamer tape has been the standard for transport of DVI® compressed video and multimedia files. With a capacity for storing 150 megabytes of data, the archive tape is being used both to transport digital data from a DVI® compression facility to the application developer and to transport one or more cartridges representing the final application to a CD-ROM pressing facility. The archive tape backup systems cost about $1,200 with a media cost for 150 megabyte cartridge of $25.

DAT TAPE

DAT tape for the storage of computer data has recently become an alternative. DAT data tape drives priced at between $1,000 and $2,000 are now available. The DAT tape format stores between one and two gigabytes of data on a small tape cartridge. The data is stored contiguously on a $35 streamer tape. The low cost of the DAT tape makes it an ideal medium for storing an entire finished application and all of its working files. The DAT tape is also ideal for exchanging data between production sites where the data can be copied back to a hard drive for playback of an application. Like the other formats, many CD-ROM pressing facilities accept DAT tape for CD-ROM production.

BERNOULLI REMOVABLE CARTRIDGE SYSTEMS

The Bernoulli removable disk system by Iomega is also an alternative for application development. The Bernoulli system has a substantial installed base and includes utilities to allow data to be easily transported between PC and Macintosh multimedia development systems.

The Bernoulli data cartridges have the capacity for storing 44 megabytes of data. During development, this can be used to store graphic or audio files and easily move them between workstations during application development. The 44-megabyte cartridge size is also ideal to store off-line working files of various elements of a production. Entire applications, including motion video, can be played back directly from the Bernoulli system adding to their convenience. Single-unit internal Bernoulli drive costs about $800. The 44M removable cartridges are $50 to $60 each.

FLOPPY DISKS

While floppy disks may not be your first choice for distribution of multimedia applications, they are a viable alternative, especially when combined with shareware file compression utilities such as ARC or PK ZIP. These public-domain software compression programs can squeeze files an additional 25 percent to 30 percent of their original size or more for multimedia files. This can be significant when you consider that an application consisting of 15 megabytes of data can be contained on 10 floppy disks.

The key advantage to the floppy disk is that every computer has one. With careful design and an eye toward conserving space, an interesting information or training application featuring stills, audio, and very short clips of motion video can be delivered on low-cost floppy disks for playback from each computer's hard disk system.

Floppy disks may become more attractive as their capacity continues to increase. The industry is already beginning to test low-cost floppy disks capable of holding over 20 megabytes of data, making them very practical for distribution of multimedia information to be played back from hard disk.

EFFICIENT TRANSPORTATION OF LARGE FILES IS A REALITY

About the same time as the industry has developed lower-cost, high-performance methods of transporting large digital data files, multimedia has emerged to take advantage of this additional capacity. The multimedia application developer has a number of options available for the distribution of multimedia data. The final choice can be made based upon the size of the intended distribution, cost of the media, security factors, and the installed base of playback systems.

Chapter

16

Copyright and the Multimedia Producer

WHAT IS A COPYRIGHT?

Copyright is your friend. Picture this. A year from now, after you have successfully produced a multimedia application that was the result of your creative efforts and financial risks, you will attend a trade show where elements of your production are reproduced in a new application that is being given away as a part of a promotion by another software producer. Your work—reproduced without your permission.

There are two ways to look at the copyright law and how it affects the production of multimedia. One way is to see it as an obstacle or a hurdle that you must overcome. Certainly, the work that is involved in obtaining rights for photographs, motion video, and text-based information can seem like an overwhelming task. On the other hand, the copyright laws were designed to protect artists, photographers, and others who make their livelihood from creating, formatting, and publishing information. This includes you, as a multimedia producer.

Copyright law is described in detail in the legal documents and is covered under federal law of the United States. Copyright law is not limited to the United States, however. Copyright is also acknowledged internationally. Basically, copyright gives the owners the right to use the materials produced and to give others the right to use those materials. This includes the right to:

Reproduce works

Create derivative works

Distribute copies

Perform the work

Display the work publicly

If you create a work and own the copyright, you can copy it, transform it, sell it, and show it. Others cannot.

Key points for multimedia

Copyright protection comes into being the moment a work is in fixed form—for multimedia that means that as pieces of the work evolve, they are protected. This is another good reason to document the process of application development. Storyboards, project plans, treatments, scripts, and flow charts are all works that can potentially need protection at some point or another. This is particularly important if you are presenting proposals to others publicly, or distributing your ideas as a way to obtain funding or a publishing contract.

If you are doing work for someone else, as a contractor, employee or subcontractor, the work typically belongs to the employer unless other arrangements have been made. You may want to propose joint ownership of a work that you are creating for someone else. At the least, it is common practice for you to have the right to show this work as a part of your portfolio. Sometimes, a client's work is strictly proprietary (or classified), and you are not permitted to show the work. In these cases, you may be able to get agreement to show part of the work or to alter it, removing sensitive material so that you can show it and talk about it publicly.

If you are creating a work with another author, the work is jointly owned, and both owners have equal right to the copyright. If the material is used again for another project, both authors have rights to it. This can also vary according to a contract that may be drawn up ahead of time. If you are a subcontractor, the work is not considered joint work. Your contribution is then considered "work for hire" and is not a jointly authored piece.

Just as your work is protected, so is that of others. If you need to use a photograph or other media element in your application, you need to understand the rights of the owner of copyright from the other side, too. All of the elements that make up multimedia that you may want to use in your application are potentially protected:

Music

Literary works

Dramatic works

Pantomime

Pictures, graphics, and sculptures (including maps and charts)

Motion pictures

Sound recordings

Acquiring rights to works can be a large effort, depending on how many works you need to acquire and how specific you are about what you need. It may be easier to acquire a photograph of Richard Nixon from the late 1960s than to acquire a specific shot from a specific photographer or news agency, for example. The more general you can make your requirements, the more flexibility your staff will have in acquiring elements. In addition, if you can be general in your requirements, you will have more flexibility from the budgetary point of view. If you have a general requirement, you will usually have a more favorable position when you are negotiating a purchase of a photograph or piece of music.

BUDGETING THE ACQUISITION OF MEDIA

Everyone agrees that it can be a costly enterprise to acquire rights for any media element that you need to purchase. At the same time, part of this is due to the fact that purchasing rights for a multimedia application are still largely uncharted waters.

No one really knows how to charge for media used in an all-digital, interactive multimedia application. Most motion-video footage use is charged by the second or minute for broadcast. Each time a motion-video piece is used, the owner will be paid for the usage. If a piece is used in a multimedia encyclopedia, however, the user (viewer) may or may not ever access any particular piece. It may be used 20 times or no times, depending on the user. In addition, it is a small piece of a much larger work and is not necessarily central, as it may be in a broadcast.

The same is true of a still image. Rights and payment for use of photographs have largely been based on book and magazine usage. A photograph purchased for use in a print publication, or even a film strip, is repurchased for every use. The economies of use costs have been developed to match the economies of distribution of these media. A CD-ROM can hold thousands of still images. It is not feasible for the same fee structure to apply for multimedia.

There are several groups trying to work on the issues that this brings up for media owners. The National Geographic Society has sponsored round table discussions about this very issue, and with good reason. The Society has a photographic library of over 10 million still images, but many of them have to be renegotiated for uses other than magazine or broadcast.

Children's Television Workshop has 20 years of footage in their archives. We were able to use a great deal of it in a prototype application we created for DVI® technology a few years ago. However, we were able to use it only because this was identified as a research project, and was covered in the union contracts that were struck when the footage was originally shot. If this footage was to be used in a real application that would be sold and distributed, rights would have to be renegotiated with the Actor's Guild.

Creators of original work seem to be concerned about two things:

1. Their work will be misrepresented. In other words, the original work will be altered in a way that was not originally intended.
2. They are concerned that they are sitting in the middle of a vast information goldmine. Photographers (like computer programmers and application developers) want to be treated fairly and to benefit financially from the creative genius and technical skill of photography that they have developed over their lifetimes.

What are the proposals for resolution of the issues raised? Groups that have met to work on these issues have come up with a number of different proposals. One is to create a clearinghouse for any material used for multimedia similar to the way ASCAP or BMI act as a clearinghouse for the music and record industry. A second is to create tiers of pricing, so that if you are distributing a CD-ROM with 5000 stills, you are not paying for each still, but for a percentage of the still that represents how often it is likely to be viewed. Until the issue of what is fair and reasonable compensation is resolved, however, you must negotiate for each still image, audio, or motion-video clip you want to use.

HOW TO ACQUIRE RIGHTS TO MEDIA

"Stock houses" specialize in selling audio, video, and motion-video clips for general use. You buy footage for certain uses, and then you are clear in terms of copyright. In addition, there are a number of application developers for DVI® technology who are specializing in buying rights to images and sounds, converting them into a format for DVI® applications, and selling CD-ROMs as a sort of "multimedia clip art."

As mentioned earlier, one of the most important things to keep in mind when designing your application is *not* to identify or use specific content prior to getting clearance. This will make your task much more difficult. Remember also that just because a still image appears in a book, or a motion-video clip runs on a particular broadcast, it is not clear that the publisher or network owns the copyright. The media may have been purchased from the owner for this one instance.

One example of this occurred during the production of the multimedia prototype, Palenque. We used an image from a book as a stand-in, and it became important to acquire the rights to show the prototype publicly. When we went to get clearance for the photograph from the publisher, we found that the copyright was retained by the photographer. It took many months to find the owner, who was living in the rain forest in Brazil. Finding him was only one half of the challenge, however. Explaining how the photograph would be used, and coming up with a fair price, was the real challenge. The mistake we made was getting stuck on one image—in hindsight it would have been better to go to a stock photography house and buy an image of a howler monkey that fit the application rather than to try to get the rights to this specific image.

Another source for material is images that are in the public domain. Images that are owned by the U.S. Government, like NASA footage, are in the public domain. Other available works include those that revert into the public domain after the author has been deceased for 50 years. This is only the case when the estate of the author does not take control of the work.

Another alternative is to create the work yourself or to contract for it. In some cases, this may be a more cost-effective alternative. Making the decision to create original works will require you to do a cost/benefit analysis that is specific to your application.

COPYRIGHT VIOLATIONS

The copyright law of the United States is very clear about infringement of law. If you copy a work, you will owe the owner payment. If someone copies your work, you will be owed money. The amount of money involved in the settlement of copyright infringement can vary, depending on whether the infringer can prove that the work was used with the intention of infringement, or without intent.

If someone uses a work that you have created, you are entitled to a minimum of $500 to $20,000 for each work, if the infringer can prove it was not intentional. If the infringer knowingly uses the work, the stakes are higher. In this instance, they owe you up to $100,000 for each work. The infringer will also probably owe you court costs, too.

If each work means each still image, each motion-video clip, and each piece of music—you stand to gain or lose a lot, depending on which side of this transaction you are on.

RESOURCES FOR COPYRIGHT PROTECTION

This short chapter is meant to raise your awareness level of copyright issues, not to teach you all there is to know about copyright protection. In fact, we probably do not know it all. Our best advice: *Make sure you are protected by having legal counsel.*

A wonderful and thorough guide to copyright protection, including sample contracts, letters, and pointers to more resources is provided in *The Multimedia Producer's Legal Survival Guide* by Stephen Ian McIntosh. This manual, and the diskettes that accompany it, can be obtained from:

Multimedia Computing Corp.
2900 Gordon Avenue, Suite 100
Santa Clara, CA 95051
(The guide is copyrighted 1990.)

Chapter

17

The Future of DVI® Technology Applications

The development of DVI® technology has made the personal computer a complete communications tool. The data types that the PC can now store, distribute, sort, retrieve, and display include the full spectrum of media. Yet the work is not complete. Companies and organizations around the world have become involved in the future of all-digital multimedia.

Some of this work will affect you as a developer in the short term. The i750 video processor may soon be integrated onto the motherboard of popular computers; software tools and application tools will continue to be developed; system software will be developed to make DVI® applications run in different operating systems like OS/2, Windows, the Macintosh, and Unix; microcode software will be written to make the i750 video processors run in conjunction with a variety of CPUs.

We would like to speculate here on long-term developments. What are the major trends in multimedia and computing, in general, that will have an impact on your work over the next four to five years? There are already a number of new advancements in several key areas including:

- International standards committees for digital stills and motion video compression

- The development of faster, more capable (and backward-compatible) video processors to improve the quality of compressed, digital video
- The integration of multimedia computing and video teleconferencing
- New technology for storing multimedia data on low-cost, rewriteable, high-performance optical disks
- The development of new fiber-optic digital computer networks with the ability to quickly and efficiently transport large amounts of data

Why are these developments important to you? One important reason to look to the future is to understand how people will be using your applications. Based on current trends, it appears sure that the cost of providing multimedia technology on personal computers will continue to drop. Many analysts in the industry have already assumed that the multimedia capabilities of presenting still images, sound, and motion video will become a standard part of the architecture of every personal computer made.

Second, as a producer of multimedia information, part of the design preproduction phase is to understand where your particular product will be in two, three, or five years. For example, if you had to produce a multimedia quarterly report for a corporation, you know that the life span of that production is only going to be until the next quarter's report production is completed. But if you have been asked to produce an educational program, it is conceivable that, like a textbook, it will have a life span of eight to ten years. With this in mind, you need to look ahead to think about how a production will be viewed five or eight years from now. There may be elements of your design and production that will vary based upon future availability of hardware products and software tools.

Let's take a look at the individual areas to see how the future of multimedia will be evolving.

STANDARDS

Developing standards for the personal computer market traditionally has not been a planned or organized process. Typically, standards have been set by one or two manufacturers, and developers have to choose to create software for one platform or another, or to create for multiple platforms. End users typically choose only one standard.

An optimistic sign of the future of multimedia is that international standard-setting organizations have already been meeting and working to implement standards which will allow applications and file formats produced on one platform to have cross compatibility on others.

Joint Photographic Expert Group—JPEG

During the late 1980s, an international standard-setting organization, ISO, began forming committees to set standards for the emerging electronic multimedia. They determined that the growth of these industries would be accelerated if all of the manufacturers agreed on common data-storage formats for still images and motion video.

The first standards-setting group was JPEG, the *Joint Photographic Experts Group*. The JPEG standards committee not only recommended a file format, but went one step further and defined an image compression approach.

The interesting result of the JPEG specification is the broad-scale support on many different computer platforms by companies in a variety of industries.

Semiconductor companies have also begun development of JPEG-compatible products. Intel supports the compression and decompression of JPEG images in products based upon the i750B video processor. The result is very good quality images at a compression ratio of 60 to 1 or less than 9 kilobytes for a typical 512 × 480, 16-bit image.

Motion Video Expert Group—MPEG

Another committee has been working simultaneously to address the compression and decompression of motion video. This group is called the *Motion Picture Experts Group,* or MPEG. The MPEG committee has been working on the complicated task of defining the standards for the compression and decompression of motion video with digital audio. Results of this committee's work are expected in the 1992-to-1993 time period. The results, like JPEG, will mean there will be one proposed standard for numerous manufacturers to follow for the compression and decompression of motion video.

One interesting goal of the MPEG committee is for good quality, efficient compression. This means that compression can be done directly on the desktop in real time, the approach taken by Intel's RTV technology. The MPEG group has determined that a resolution of 352 × 240 is optimal for the standard to meet CD-ROM data rates.

High-quality audio is also a key part of the MPEG standard. The early work of the MPEG committee has determined that an ADPCM algorithm should be used (the same as with PLV and RTV). The MPEG standard will also specify the best method for interleaving the digital audio stream together with digital video.

POWERFUL NEW PROGRAMMABLE VIDEO PROCESSORS

The second development on the horizon for multimedia applications is a high-performance video processor necessary to implement higher-quality real-time compression and decompression of motion video. A screen full of data changing at 30 frames per second is a formidable task.

For most of the computing tasks, the computing power available with today's 286 and 386 microprocessors has reached optimal efficiency for personal computer application software computations. To put this more succinctly, we are beyond the point where adding computing power to a word processor, the average spreadsheet, or to read our e-mail, will make a big difference. Now, the advancements in microprocessor speed and performance are being dedicated toward making computers easier to use by offering pictures and sound to enhance the communications and information transmission capability.

Gordon Moore, one of the founders of Intel, long ago had the foresight to see that microprocessor design and manufacturing technology would allow the power of the microprocessor to more than double every two years. While this processing power doubles, the size of the high-performance microprocessor continues to decrease, consequently lowering manufacturing cost and the cost of this added performance.

The emergence of worldwide standards now gives the software and microprocessor designers a direction to focus this processing power. Today, DVI® technology processing capabilities are handled by a set of video processors separate from the host microprocessor.

The lessons learned by Intel and the design philosophy behind its ×86 architecture are not lost on the video processor, however. Intel's solution to a business environment where standards are only now emerging, and are likely to change, is to create a series of video processors that are programmable. Like the Intel386 and Intel486 microprocessors, the data created today will be able to be played on future versions of i750 video processors. As MPEG standards are announced, a future i750 processor will be capable of handling the intensive amount of data and processing power needed to play these motion video files. And as the video processor is merged with the central processor, the investment of application developers is protected.

By putting all of this capability onto one chip, the manufacturing cost of the multimedia personal computer will continue to drop. Performance will also increase as the component's power increases. This decrease in cost and its associated decrease in physical size indicates the potential for very small, low-cost, multimedia personal computers. Couple this with the emergence affordable, writeable CD-ROM technology, and we begin to see the indicators of what the future of personal computers and communications can be.

THE EMERGENCE OF HIGHER-CAPACITY CD-ROM DEVICES

The development of laser technology for data storage continues to improve and advance. Three years ago, the thought of a $500 device capable of storing 650 megabytes of removable data disks, each with a manufacturing cost of less than $2.00, was staggering. When it was first developed, software developers did not completely understand how all of this data-storage capability would be used. The increasing use of multimedia answered the question.

Industry speculation has the next generation of CD-ROMs set at quadruple the data rate and storage capability. This is sometimes called *quad-density quad-data rate CD-ROMs.* While today's CD-ROMs will hold up to 72 minutes of motion video, tomorrow's CD-ROMs will probably trade off data storage for physical size, reducing the disk size to a three-inch standard.

Quad-data rate combined with more efficient video processors and industry standard algorithms present the opportunity to use this capacity in a number of ways:

There is potential to double, and even triple, the quality of today's motion video display.

The ability to mix effects, animation, and graphics in the middle of the motion-video display can be developed.

Sheer volume of information per CD-ROM can be increased.

The ideas are endless, and will be driven by developers like yourself.

Rewriteable disks

Rewriteable technology is available today, but costs several thousand dollars for a 500-megabyte to 1-gigabyte removable-cartridge optical-storage system. As we have seen with other technologies, costs are already beginning to drop steadily, and it appears that rewriteable optical storage will be available in the very near future for costs approaching today's CD-ROM prices.

This development opens up the ability to easily transport large amounts of data in situations where high-capacity local area network is neither available nor efficient.

ADVANCEMENTS IN LANS TO TRANSPORT MULTIMEDIA DATA

While it is a convenience to be able to share files and data, the principle driving force behind the adoption of Local Area Networks (LANs) and Wide Area Networks (WANs) is their desire to communicate quickly and easily through e-mail. It is very clear that the addition of multimedia capabilities on the personal computer, along with advances in the speed and capacity of local area networks, will allow the multimedia personal computer users to communicate with enriched e-mail. Enriched e-mail will offer the user a choice of transmitting text, as they do today, with the addition of pictures, graphics, voice mail, and even motion video.

Multimedia databases

The development of the laser is also creating new ways to distribute large amounts of information over *Fiber Optic Distributed Data Information,* or FDDI. This advance promises to significantly increase today's local area network data transmission rate by a factor of 10. The result will be information that you can actually see and hear, accessible from mass storage.

It is fascinating to think of the possibilities of having the video archives of Cable News Network on-line for access, or The National Geographic Society's hours and hours of video, film, and audio information accessible in our homes, schools, and at work at the touch of a key.

How far away is the future?

All of the future technologies discussed in this chapter are demonstrable today. Companies are working to merge the high-performance capabilities of microprocessors, optical storage, and high-speed data transmission together to complete what Bill Gates, the chairman and founder of Microsoft, calls a world of "information at our fingertips." It appears highly likely that the advancements presented in this chapter will be a reality within the next three to four years.

SOMETHING IS MISSING

The missing ingredient to making all of the above technical advancements meaningful is the development of the software applications. The ability to have "information at our fingertips" has been the dream of educators, corporate executives, advertising designers, retailers, travel specialists, computer analysts—the list is exhaustive. The foundation for the technology is already in place. The opportunity is for content providers and producers, such as yourself, who will learn this technology inside and out and utilize it to fill the needs of business and education around the world. The opportunity to take part in this evolution of communications, publishing, and personal computers is yours.

Appendix

Additional Reading

Multimedia References

The CD-ROM Yearbook 1989–1990
Microsoft Press
(March 1989)
ISBN 1-55615-179-9

Digital Video in the PC Environment
Arch C. Luther
McGraw-Hill (1990)
ISBN 0-07-039176-9

Interactive Multimedia
Visions of Multimedia for Developers, Educators, and Information Providers
Edited by Sueann Ambron and Kristina Hooper
Microsoft Press (1988)
ISBN 1-55615-124-7

Managing Interactive Video/Multimedia Projects
By Robert Bergman and Thomas Moore
Educational Technology Publications (1990)
ISBN 0-87778-209-1

The Multimedia Producer's Legal Survival Guide
Stephen McIntosh
Multimedia Publishing Corp. (1990)
ISMBN 1-878955-00-4

On Multimedia
Technologies for the 21st Century
Edited by Martin Greenberger
The Voyager Company
ISBN 1-55940-141-9

Other Readings of Interest

Making CBT Happen
Gloria Gery
Weingarten Publications (1987)
ISBN 0-9617968-0-4

The Media Lab
Inventing the Future at MIT
Stewart Brand
Viking Penguin, Inc. (1987)
ISBN 0-670-81442-3

The Mind's New Science
A History of the Cognitive Revolution
Howard Gardner
Basic Books
ISBN 0-465-0463505

Appendix

Resources for Multimedia Production

The following companies are sources for many of the software and hardware products, and services mentioned in this book.

For information on ActionMedia products:

Intel
3065 Bowes
Santa Clara, CA 95051
(800) 548-4725

IBM
Marketing Services
P.O. Box 2150
Atlanta, GA 30301–2150
(800) 344-3155

DVI® TECHNOLOGY VAR/SYSTEM INTEGRATOR

One complete source for systems, software, and peripherals.

Avtex Research Corporation
2105 S. Bascom Avenue, Suite 290
Campbell, CA 95008
(408) 371-2800

AUTHORING SOFTWARE

Authology:MultiMedia
CEIT Systems, Inc.
4800 Great America Parkway
San Jose, CA 95054
(408) 986-1101

MEDIAscript
Network Technology Corporation
7401-F Fullerton Road
Springfield, VA 22153–3122
(703) 866-9000

Hyperties for DVI® Technology
Cognetics
55 Princeton Heights Town Road
Princeton Junction, NJ 08550
(609) 799-5005

GRAPHICS SOFTWARE

Lumena for DVI® Technology
Time Arts
3436 Mendocino
Santa Rosa, CA 95403
(707) 576-7722

VIDEO EDITING SOFTWARE

D/Vision
TouchVision Systems Inc.
1800 Winnemac Ave.
Chicago, IL 60640
(312) 989-2160

GRAPHICS BOARDS

Truevision
7340 Shadeland Station
Indianapolis, IN 46256
(317) 841-0332

PRINTERS

Canon U.S.A. Inc.
One Canon Plaza
Lake Success, NY 11042
(516) 488-6700

Sony Corporation of America
Sony Drive
Park Ridge, NJ 07656
(201) 930-1000

DIGITIZING TABLETS

WACOM Inc. West
115 Century Road
Parmus, NJ 07652
(201) 265-4226

CD-ROMS

Sony Corporation of America
Optical Memory Group
655 River Oaks Parkway
San Jose, CA 95134
(408) 432-0190

FLATBED SCANNERS

Howtek, Inc.
21 Park Avenue
Hudson, NH 03051
(603) 882-5200

Sony Corporation of America
Sony Drive
Park Ridge, NJ 07656
(201) 930-1000

SLIDE SCANNERS

Nikon Inc.
1300 Walt Whitman Road
Melville, NY 11747-3064
(516) 547-4355

VIDEO EQUIPMENT

Video Cameras

JVC Professional Products Company
Division of JVC Corp
41 Slater Drive
Elmwood Park, NJ 07407
(800) 582-5825

Time Base Corrector

Sony Corporation of America
Sony Drive
Park Ridge, NJ 07656
(201) 930-1000

AUDIO PRODUCTION EQUIPMENT

TEAC America, Inc.
7733 Telegraph Road
Montebello, CA 90640
(213) 726-0303

Roland Corp US
7200 Dominion Circle
Los Angeles, CA 90040
(213) 685-5141

STOCK MUSIC AND SOUND EFFECTS

Creative Support Services
1950 Riverside Drive
Los Angeles, CA 90039
(213) 666-7968

DeWolfe Music/SFX Library
25 W. 45th Street
New York, NY 10036
(800) 221-6713

First Com/Music House/Chappell
13747 Montfort #220
Dallas, TX 75240
(800) 858-8880

Anchor Audio
(213) 533-5984

Promusic
(800) 322-7879

Omnimusic Production Music Library
52 Main Street
Port Washington, NY 11050
(800) 828-6664

STOCK PICTURES

The Image Bank
111 Fifth Avenue
New York, NY 10003
(212) 529-6700

Photovault
1045 17th Street
San Francisco, CA 94107
(415) 552-9682

Tony Stone Worldwide Stock Agency
233 East Ontario Street, Suite 1200
Chicago, IL 60611
(312) 787-8798

PRODUCTION PLANNING SOFTWARE

Microsoft Corporation
One Microsoft Way
Redmond, WA 98052-6399
(206) 882-8080

CD-ROM PRESSING FACILITIES

3M
1425 Parkway Drive
P.O. Box 368
Menomonie, WI 54751
(715) 235-5568

Discovery Systems
7001 Discovery Blvd.
Dublin, OH 43017
(614) 761-2000

LEGAL GUIDE

Stephen Ian McIntosh, *The Multimedia Producer's Legal Survival Guide*
Available from:
Multimedia Computing Corporation
2900 Gordon Avenue, Suite 100
Santa Clara, CA 95051
(408) 737-7575

MULTIMEDIA NEWSLETTERS

Mind Over Media, Bi-Monthly Newsletter of "How To's for Multimedia Professionals"

Multimedia Computing Corporation
2900 Gordon Avenue, Suite 100
Santa Clara, CA 95051
(408) 737-7575

Multimedia Computing & Presentations
Industry News and Updates

Multimedia Computing Corporation
2900 Gordon Avenue, Suite 100
Santa Clara, CA 95051
(408) 737-7575

Multimedia & Videodisc Monitor

Future Systems, Inc.
P.O. Box 26
Falls Church, VA 22207
(703) 241-1799

INDUSTRY ORGANIZATIONS

IMA—Interactive Multimedia Association
800 K Street, NW
Suite 440
Washington, DC 20001
(202) 408-1000

Appendix

System Configurations

Lumena for DVI® technology configuration files

AUTOEXEC.BAT

```
@echo off
set vroot=C:\V
set path=%VROOT%\bin;c:\dos;c:\386max
prompt $P$G
c:\v\bin\vvid0006 -KY -CY
```

CONFIG.SYS

```
files=20
buffers=20/x
stacks=0,0
break=on
lastdrive=z
device=c:\v\bin\386max.sys pro=pro.sys
device=c:\aspi4dos.sys /D
shell=c:\command.com /p /e:1600
```

PRO.LUM

```
norom
vidmem=b000-c000
ram=d000-e000
```

Authology:MultiMedia configuration files

AUTOEXEC.BAT

```
@echo off
set vroot=C:\V
set path=%VROOT%\bin;c:\dos;c:\386max;
set vdisks=c0m w0c
set vav=%VROOT%\av
set im=%VROOT%\vid\im
prompt $P$G
c:\v\bin\vvid0006 -KY -CY
c:\v\bin\vaudam
mouse
mscdex /D:CD1 /M:10 /L:W /E
vset -DD
```

CONFIG.SYS

```
files=20
buffers=20 /x
stacks=0,0
break=on
lastdrive=z
device=c:\v\bin\vram.sys
device=c:\v\bin\vscsi001.sys
device=c:\386max\386max.sys pro=pro.sys
device=c:\aspi4dos.sys /D
shell=c:\command.com /p /e:1600
```

PRO.AAM

```
norom
vidmem=b000-c000
use=a000-b000
ram=d000-e000
```

Glossary

ActionMedia™. DVI® technology's product family, consisting of single-board delivery and single-board capture boards for AT and Micro Channel™ architecture buses. For example, ActionMedia 750 Delivery Board.

Algorithm. A set of mathematical operations that perform a particular function, like compression of an image.

Alias. One form of image distortion, commonly manifested through a stairstepped appearance on curved or diagonal lines.

Antialias. The electronic process of removing alias from an image or from text fonts.

Artifact. An unintended, unwanted visual aberration in a video image.

Aspect ratio. The ratio of the width to the height on an image. Video aspect ratio is typically 4:3.

Asymmetric compression. A compression scheme that takes more than the running time of the video to compress, resulting in a higher quality image than systems that compress video at the real-time rate of 30 frames per second. For example, a minute of video would take more than one minute to compress. Intel's Production Level Video (PLV) is one example.

Bit. The smallest unit of memory in a computer, an on or off signal represented by 1 or 0.

Byte. String of bits (usually 8 bits) to represent one data character (letter, digit).

CCITT. Consultative Committee of Telegraph and Telephone. International standards committee whose charter is to generate standards for networking and telephony.

CD-ROM (Compact Disc Read Only Memory). An optical storage media (4.75 inches in diameter) holding over 660 megabytes of digital data.

Chrominance. The color information in an image, hue, and saturation.

Composite Video. Analog systems used to transmit and record video programs, containing all color information in one signal, such as NTSC, PAL, and SECAM.

DAT (Digital Audio Tape). A consumer recording and playback media for high-quality audio.

Digitize. To convert analog data to digital data.

Discrete Cosine Transform (DCT). One from of encoding (mathematical technique or algorithm) used in image compression.

Field. One-half of a complete video frame, consisting of every other scan line.

Frame. One complete image in a motion video display. Motion is created by playing a series of frames at a prescribed speed.

Frame rate. The speed at which video images are displayed.

High Sierra format. A standard format for placing files and directories on CD-ROM, revised and adopted by the International Standards Organization as the ISO 9660.

i750™. Intel's programmable video processor enabling video compression and decompression in a personal computer.

Interlace. Scheme to display a video image by displaying alternate scan lines in two discrete fields.

International Standards Organization (ISO). The worldwide group responsible for establishing and managing the standards committees and expert groups, including those working on international image compression standards.

Interpolation. The averaging of pixels, used when scaling an image to reduce pixels or create new ones. Also referred to as "line-doubling" when an image is being scaled to a larger size.

JPEG (Joint Photographic Experts Group). An ISO group chartered with generating standards for full-color still-image compression.

Kilobytes (KB). 1024 bytes.

Lossless compression. A compression technique that preserves all the original information in an image or other data structures.

Lossy compression. A compression technique that achieves optimal data reduction by actually discarding redundant and unnecessary information in an image.

Luminance. The brightness values in an image, or all the black and white information in an image.

Megabytes (MB). One million bytes.

MIDI (Musical Instrument Digital Interface). A standard for connecting electronic instruments such as synthesizers to the personal computer.

MPEG (Motion Picture Experts Group). An ISO group chartered with generating standards for motion picture video compression from digital storage media.

Px64. Informal name for the CCITT video algorithm (also known as H.261) proposed for teleconferencing.

PIC (Picture Image Compression). One compression scheme for still images available when using DVI® technology products.

PLV (Production Level Video). DVI® technology's "off-line," or asymmetrical compression algorithm for motion video. Currently, this compression will deliver the highest quality motion video image to DVI® developers.

RGB (Red-Green-Blue). Computer color display with three distinct signals—red, green, and blue.

RTV (Real Time Video). DVI® technology's "on-line," or symmetrical compression algorithm for motion video. This algorithm is used in conjunction with ActionMedia products installed in the personal computer.

Scanning. In data capture, the process of electronically converting an image to a digital (or machine-readable) signal.

SMPTE time code. Standardized edit time code adopted by SMPTE, the Society of Motion Picture and Television Engineers.

Symmetric Compression. A compression scheme that takes an equal amount of time to compress an image sequence as it does to play that image sequence. For instance, a minute of running video would take one minute to compress using a symmetrical compression scheme. Intel's Real Time Video (RTV) is one example.

Time code. A frame-by-frame address code used in video to mark each frame by hour, minute, second, and video frame (eight digits).

WORM (Write Once/Read Many). An optical storage media that allows users to record, but not erase, data.

Index

About the Authors

Mark J. Bunzel is president of Avtex Research Corporation, a company that develops multimedia software and systems for application developers and corporate clients as NASA, Lockheed, Apple, Federal Express, IBM, and Intel Corporation. Mr. Bunzel teaches multimedia interactive video, video production planning and budgeting, and multimedia desktop video at the University of California at Santa Cruz Extension Division. He has played a major role in the strategic planning and marketing of DVI® technology.

Sandra K. Morris is the manager of Developer Programs for the market development organization of Intel's Multimedia Products Operation, a group dedicated to communicating with software developers worldwide about the potential of multimedia software applications. She has had vast experience working with DVI® technology, including application design and development, and strategic marketing planning.